THE COPENHAGEN DIAGNOSIS

UPDATING THE WORLD ON THE
LATEST CLIMATE SCIENCE

Dedication

Stephen H. Schneider (1945 - 2010)
Pioneering climate researcher and eloquent
science communicator.

THE COPENHAGEN DIAGNOSIS

UPDATING THE WORLD ON THE LATEST CLIMATE SCIENCE

ELSEVIER

AMSTERDAM • BOSTON • HEIDELBERG • LONDON
NEW YORK • OXFORD • PARIS • SAN DIEGO
SAN FRANCISCO • SINGAPORE • SYDNEY • TOKYO

Elsevier
30 Corporate Drive, Suite 400, Burlington, MA 01803, USA
525 B Street, Suite 1900, San Diego, CA 92101-4495, USA
The Boulevard, Langford Lane, Kidlington, Oxford, OX5 1GB, UK

Library of Congress Cataloging-in-Publication Data
The Copenhagen diagnosis : updating the world on the latest climate science/contributing authors, Ian Allison … [et al.].
     p. cm.
    Includes bibliographical references.
    ISBN 978-0-12-386999-9
    1. Climatic changes. I. Allison, Ian.
    QC903.C67 2011
    577.2′2—dc22
                                    2011001542

British Library Cataloguing-in-Publication Data
A catalogue record for this book is available from the British Library

ISBN: 978-0-12-386999-9

For information on all Elsevier publications visit our website at *www.elsevierdirect.com*

Printed and bound in China
11 12 13   10 9 8 7 6 5 4 3 2 1

HELENA EGE
University of Montana
LIBRARY
1115 N.Roberts
Helena, MT 59601

Working together to grow
libraries in developing countries

www.elsevier.com | www.bookaid.org | www.sabre.org

ELSEVIER   BOOK AID International   Sabre Foundation

47.45

Photo credits:
Text: page x ©Rainer Prinz *Weissbrunnferner, Italian Alps, 18 July 2006, showing a glacier that has lost its firm body. Extended dark ice surfaces accelerate the melt rate*, page xii ©evirgen & NASA - iStockphoto®, page 19 ©Luoman *Amazon rainforest deforestation*- iStockphoto®, page 37 ©Stephen Schneider *Sunset giant iceberg at Ilulissat*, page 45 ©Phil Dickson *Ice stack collapsing off the Perito Moreno Glacier, Patagonia Argentina* - iStockphoto®, page 54 ©Sebastian D'Souza *Indian commuters walk through floodwater* - Getty Images®, page 66 ©Maxim Tupikov *Arctic icebreaker* - iStockphoto®, page 68 ©Alexander Hafeman (Mlenny) *Dead Vlei Namibia* - iStockphoto®, page 69 ©E. Steig, page 71 ©Ian Joughin *Meltwater on the Greenland Ice Sheet*, page 94 Muammer Mujdat Uzel *Marl and dry land on recent lake Denizili Turkey* - iStockphoto®, page 99 ©Kirill Putchenko - iStockphoto®.

Contributing Authors

Ian Allison
Nathaniel Bindoff
Robert Bindschadler
Peter Cox
Nathalie de Noblet-Ducoudré
Matthew England
Jane Francis
Nicolas Gruber
Alan Haywood
David Karoly
Georg Kaser
Corinne Le Quéré
Tim Lenton
Michael Mann
Ben McNeil
Andy Pitman
Stefan Rahmstorf
Eric Rignot
Hans Joachim Schellnhuber
Stephen Schneider
Steven Sherwood
Richard Somerville
Konrad Steffen
Eric Steig
Martin Visbeck
Andrew Weaver

Contents

Preface

It is over four years since the drafting of text was completed for the Intergovernmental Panel on Climate Change (IPCC) Fourth Assessment Report (AR4). In the meantime, many thousands of papers have been published on a suite of topics related to human-induced climate change. The purpose of this book is to synthesize the most policy-relevant climate science published since the close-off of material for the last IPCC report. The rationale is two-fold. Firstly, this report serves as an interim evaluation of the evolving science midway through an IPCC cycle—IPCC AR5 is not due for completion until 2013. Secondly, and most importantly, the report served as a handbook of science updates that supplements the IPCC AR4 in time for Copenhagen in December, 2009.

This report covers the range of topics evaluated by Working Group I of the IPCC, namely the Physical Science Basis. This includes:

- an analysis of greenhouse gas emissions and their atmospheric concentrations, as well as the global carbon cycle;
- coverage of the atmosphere, the land surface, the oceans, and all of the major components of the cryosphere (land ice, glaciers, ice shelves, sea ice, and permafrost);
- paleoclimate, extreme events, sea level, future projections, abrupt change, and tipping points;
- separate boxes devoted to explaining some of the common misconceptions surrounding climate change science.

The report has been purposefully written with a target readership of policy-makers, stakeholders, the media, and the broader public. The science it contains is based on the most credible and significant peer-reviewed literature available at the time of publication. The authors primarily comprise previous IPCC lead authors familiar with the rigor and completeness required for a scientific assessment of the climate science.

This report has been distributed to all national greenhouse and climate change departments and has also been made freely available on the web at:

www.copenhagendiagnosis.org

Acknowledgement

The authors wish to thank Mr Stephen Gray of the UNSW Climate Change Research Centre (CCRC) who provided invaluable logistical support and editorial assistance.

Executive Summary

The most significant recent climate change findings are:

Surging greenhouse gas emissions: Global carbon dioxide emissions from fossil fuels in 2008 were 40% higher than those in 1990. Even if global carbon emission rates are stabilized at present-day levels, just 20 more years of emissions would give a 25% probability that warming exceeds 2 °C, even with zero emissions after 2030. Every year of delayed action increases the chances of exceeding 2 °C warming.

Recent global temperatures demonstrate human-induced warming: Over the past 25 years temperatures have increased at a rate of 0.19 °C per decade, in very good agreement with predictions based on greenhouse gas increases. Even over the past 10 years, despite a decrease in solar forcing, the trend continues to be one of warming. Natural, short-term fluctuations are occurring as usual, but there have been no significant changes in the underlying warming trend.

Acceleration of melting on ice sheets, mountain glaciers, and ice caps: A wide array of satellite and ice measurements now demonstrate beyond doubt that both the Greenland and Antarctic ice sheets are losing mass at an increasing rate. Melting of mountain glaciers and ice caps in other parts of the world has also accelerated since 1990.

Rapid Arctic sea ice decline: Summer-time melting of Arctic sea ice has accelerated far beyond the expectations of climate models. The area of summer-time sea ice during 2007–2009 was about 40% less than the average prediction from IPCC AR4 climate models.

Current sea level rise underestimated: Satellites show recent global average sea level rise (3.4mm/yr over the past 15 years) to be ~80% above past IPCC predictions. This acceleration in sea level rise is consistent with a doubling in contribution from melting of glaciers, ice caps, and the Greenland and West Antarctic ice sheets.

Sea level predictions revised: By 2100, global sea level rise is likely to increase at least twice as much as projected by Working Group 1 of the IPCC AR4; for unmitigated emissions it may well exceed 1 m. The upper limit has been estimated as ~2 meters sea level rise by 2100. Sea level will continue to rise for centuries after global temperatures have been stabilized, and several meters of sea level rise must be expected over the next few centuries.

Delay in action risks irreversible damage: Several vulnerable elements in the climate system (e.g., continental ice sheets, Amazon rain forest,

West African and Indian monsoons and others) could be pushed toward abrupt or irreversible change if warming continues in a business-as-usual way throughout this century. The risk of transgressing critical thresholds ("tipping points") increases strongly with ongoing climate change. Thus waiting for higher levels of scientific certainty could mean that some tipping points will be crossed before they are recognized.

The turning point must come soon: If global warming is to be limited to a maximum of 2 °C above pre-industrial, global emissions need to peak between 2015 and 2020 and then decline rapidly. To stabilize climate, a decarbonized global society—with near-zero emissions of CO_2 and other long-lived greenhouse gases—needs to be reached well within this century. More specifically, the average annual per-capita emissions will have to shrink to well under 1 ton CO_2 by 2050. This is 80–95% below the per-capita emissions in developed nations in 2000.

1

Greenhouse Gases and the Carbon Cycle

KEY POINTS

- Global carbon dioxide (CO_2) emissions from fossil fuel burning in 2008 were nearly 40% higher than those in 1990, with a 2.5-fold acceleration over the past 20 years.

- Global CO_2 emissions from fossil fuel burning are tracking near the highest scenarios considered so far by the Intergovernmental Panel on Climate Change (IPCC).

- The fraction of CO_2 emissions absorbed by the land and ocean CO_2 reservoirs has likely decreased by ~5% (from 60 to 55%) in the past 50 years, though uncertainty is large.

GLOBAL CARBON DIOXIDE EMISSIONS

In 2008, combined global emissions of CO_2 from fossil fuel burning, cement production, and land use change (mainly deforestation) were 24% higher than the year 1990 (Le Quéré *et al.*, 2009). Of this combined total, the CO_2 emissions from fossil fuel burning and cement production were 38% higher in 2008 compared to 1990. The global rate of fossil fuel CO_2 emissions has accelerated over the last 20 years, increasing from 1.0% per year in the 1990s to 2.5% per year between 2000 and 2009 (Fig. 1.1). The accelerated growth in fossil fuel CO_2 emissions since 2000 was primarily caused by fast growth rates in emerging countries (particularly China) in part due to increased international trade of goods (Peters & Hertwich, 2008), and an increasing reliance on coal as a fuel source (Raupach *et al.*, 2007). The observed acceleration in fossil fuel CO_2 emissions is tracking the upper-end of the emissions scenarios used by IPCC AR4 (Nakicenovic *et al.*, 2000). In contrast, CO_2 emissions from

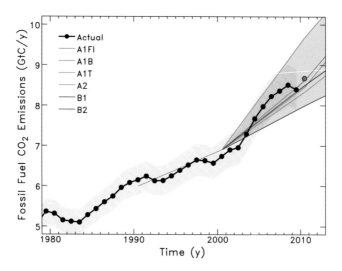

FIGURE 1.1   Observed global CO_2 emissions from fossil fuel burning and cement production compared with IPCC emissions scenarios used by the IPCC to project climate change. Observations are from the US Department of Energy Carbon Dioxide Information Analysis Center (CDIAC) up to 2007. Data for the year 2008 and 2009 are based on BP economic data. Emissions in 2010 (shown as a red point on graph) are projected to increase by more than 3%. *Data from Friedlingstein et al. (2010).* Adapted from Nature GeoScience *Vol 2, Le Quéré et al, "Trends in the sources and sinks of carbon dioxide" p. 831–836, Fig 1a, copyright 2009*

land use change were relatively constant in the past few decades, and may have decreased since 2000 (Friedlingstein *et al.*, 2010). Total CO_2 emissions have dropped by 1.9% in 2009 as a result of the global financial recession, but they are projected to increase again by more than 3% in 2010 (Fig. 1.1).

Carbon Dioxide

The concentration of CO_2 in the atmosphere reached 389 parts per million (ppm) in 2010 (Fig. 1.2). The atmospheric CO_2 concentration is more than 105 ppm above its natural pre-industrial level. The present concentration is higher than at any time in the last 800,000 years, and potentially the last 3 to 20 million years (Lüthi *et al.*, 2008; Raymo *et al.*, 1996; Tripati *et al.*, 2009). CO_2 levels increased at a rate of 1.9 ppm/year between 2000 and 2009, compared to 1.5 ppm/year in the 1990s. This rate of increase of atmospheric CO_2 is more than 10 times faster than the highest rate that has been detected in ice core data; such high rates would be discernable in ice cores if they had occurred at any time in the last 22,000 years (Joos & Spahni, 2008).

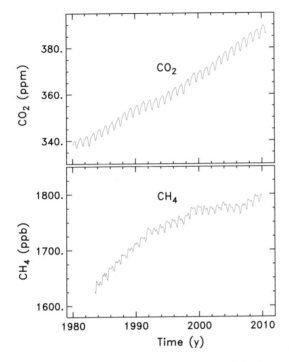

FIGURE 1.2   Concentration of CO_2 (top) and CH_4 (bottom) in the atmosphere. CO_2 and CH_4 are the two most important anthropogenic greenhouse gases. Data are global averages from the Earth System Research Laboratory of the US National Oceanic and Atmospheric Administration, Global Monitoring Division.

Methane

The concentration of methane (CH_4) in the atmosphere increased since 2007 to 1800 parts per billion (ppb) after almost a decade of little change (Fig. 1.2). The causes of the recent increase in CH_4 have not yet been determined. The spatial distribution of the CH_4 increase shows that an increase in Northern Hemisphere CH_4 emissions has played a role and could dominate the signal (Rigby *et al.*, 2008), but the source of the increase is unknown. CH_4 is emitted by many industrial processes (ruminant farming, rice agriculture, biomass burning, coal mining, and gas and oil industry) and by natural reservoirs (wetlands, permafrost, and peatlands). Annual industrial emissions of CH_4 are not available as they are difficult to quantify. CH_4 emissions from natural reservoirs can increase under warming conditions. This has been observed from permafrost thawing in Sweden (see Chapter 5), but no large-scale evidence is available to clearly connect this process to the recent CH_4 increase. If the CH_4 increase is caused by the response of natural reservoirs to warming, it could continue for decades to centuries and enhance the greenhouse gas burden of the atmosphere.

CARBON SINKS AND FUTURE VULNERABILITIES

The oceanic and terrestrial CO_2 reservoirs — the "CO_2 sinks"— have continued to absorb more than half of the total emissions of CO_2. However, the fraction of emissions absorbed by the reservoirs has likely decreased by ~5% (from 60 to 55%) in the past 50 years (Canadell et al., 2007). The uncertainty in this estimate is large because of the significant background interannual variability and because of uncertainty in CO_2 emissions from land use change.

Models show that the response of the land and ocean CO_2 sinks to climate variability and recent climate change can account for the decrease in uptake efficiency of the sinks suggested by the observations (Le Quéré et al., 2009). A long-term decrease in the efficiency of the land and ocean CO_2 sinks would enhance climate change via an increase in the amount of CO_2 remaining in the atmosphere. Many new observational studies have shown a recent decrease in the efficiency of the oceanic carbon sink at removing anthropogenic CO_2 from the atmosphere in large regions of the world. In the Southern Ocean, the CO_2 sink has not increased since 1981 in spite of the large increase in atmospheric CO_2 (Le Quéré et al., 2007; Metzl, 2009; Takahashi et al., 2009). The Southern Ocean trends have been attributed to an increase in winds, itself a likely consequence of ozone depletion (Lovenduski et al., 2008). Similarly, in the North Atlantic, the CO_2 sink decreased by ~50% since 1990 (Schuster et al., 2009), though part of the decrease has been associated with natural variability (Thomas et al., 2008). As yet, there is no direct evidence of large-scale changes in the efficiency of the terrestrial sink.

Future vulnerabilities of the global CO_2 sinks (ocean and land) have not been revised since the IPCC AR4. Our current understanding indicates that the natural CO_2 sinks will decrease in efficiency during this century, and the terrestrial sink could even start to emit CO_2 (Friedlingstein et al., 2006). The response of the sinks to elevated CO_2 and climate change is shown in models to amplify global warming by 5–30%. The observations available so far are insufficient to provide greater certainty, but they do not exclude the largest global warming amplification projected by the models (Le Quéré et al., 2009).

Q & A

Is the Greenhouse Effect Already Saturated, so that Adding more CO_2 Makes no Difference?

No, not even remotely. It isn't even saturated on the runaway greenhouse planet Venus, with its atmosphere made up of 96% CO_2 and a surface temperature of 467 °C, hotter even than Mercury (Weart and Pierrehumbert, 2007). The reason is simple: the air gets ever thinner when we go up higher in the atmosphere. Heat radiation escaping into space mostly occurs higher up in the atmosphere, not at the surface — on average from an altitude of about 5.5 km. It is here that adding more CO_2 *does* make a difference. When we add more CO_2, the layer near the surface where the CO_2 effect is largely saturated gets thicker — one can visualize this as a layer of fog, visible only in the infrared. When this "fog layer" gets thicker, radiation can only escape to space from higher up in the atmosphere, and the radiative equilibrium temperature of −18 °C therefore also occurs higher up. That upward shift heats the surface, because the temperature increases by 6.5 °C per kilometer as one goes down through the atmosphere due to the pressure increase. Thus, adding one kilometer to the "CO_2 fog layer" that envelopes our Earth will heat the surface climate by about 6.5 °C.

2

The Atmosphere

KEY POINTS

- Global air temperature, humidity, and rainfall trend patterns exhibit a distinct fingerprint that cannot be explained by phenomena apart from increased atmospheric greenhouse gas concentrations.

- Continuing the overall warming trend of recent decades, and despite natural year-to-year variations, the decade of 2001—2010 was significantly warmer on global average than that of 1991—2000, which in turn was significantly warmer than that of 1981—1990.

- Global atmospheric temperatures thus maintain a strong warming trend since the 1970s (~0.6 °C), consistent with expectations of greenhouse induced warming.

GLOBAL TEMPERATURE TRENDS

IPCC AR4 presented "an unambiguous picture of the ongoing warming of the climate system." The atmospheric warming trend continues to climb despite 2008 being cooler than 2007 (Fig. 2.1). For example, the IPCC gave the 25-year trend as 0.177 ± 0.052 °C per decade for the period ending 2006 (based on the HadCRUT data). Updating this by including 2007 and 2008, the trend becomes 0.187 ± 0.052 °C per decade for the period ending 2008. The recent observed climate trend is thus one of ongoing warming, in line with IPCC predictions.

Year-to-year differences in global average temperatures are unimportant in evaluating long-term climate trends. During the warming observed over the 20th century, individual years lie above or below the long-term trend line due to internal climate variability (like 1998); this is a normal and natural phenomenon. For example, in 2008 a La Niña occurred, a climate pattern which naturally causes a temporary dip in the average global temperature. At the same time, solar output was also at its lowest level of the satellite era, another temporary cooling influence.

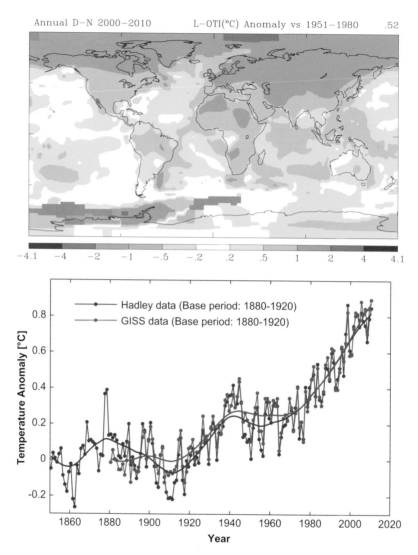

FIGURE 2.1 (top) Mean surface temperature change (°C) for 2000–2010 relative to the baseline period of 1951–1980 and (bottom) global average temperature 1850–2010 relative to the baseline period 1880–1920 estimated from the (top) NASA/GISS data set and (bottom) NASA/GISS and Hadley data. Data from the NOAA reconstructed sea surface temperature show similar results.

Without anthropogenic warming these two factors should have resulted in the 2008 temperature being among the coolest in the instrumental era, while in fact 2008 was the 9th warmest on record. This underpins the strong greenhouse warming that has occurred in the atmosphere over the past century. The most recent 10-year period is warmer than the previous

10-year period, and the longer-term warming trend is clear and unambiguous (Fig. 2.1).

IS THE WARMING NATURAL OR HUMAN INDUCED?

Our understanding of the causes of the recent century-scale trend has improved further since the IPCC AR4. By far the greatest part of the observed century-scale warming is due to human factors. For example, Lean and Rind (2008) analyzed the role of natural factors (e.g., solar variability, volcanoes) versus human influences on temperatures since 1889. They found that the sun contributed only about 10% of surface warming in the last century and a negligible amount in the last quarter century, less than in earlier assessments. No scientific literature has been published since the AR4 assessment that supports alternative hypotheses to explain the warming trend.

IS WARMING OCCURRING HIGHER UP IN THE ATMOSPHERE?

The IPCC AR4 noted a remaining uncertainty in temperature trends in the atmosphere above the lowest layers near the Earth's surface. Most data sets available at that time showed weaker than expected warming in the atmospheric region referred to as the tropical upper troposphere, 10 to 15 km above the surface. However, the observations suffered from significant stability issues especially in this altitude region. Researchers have since performed additional analyses of the same data using more rigorous techniques, and developed a new method of assessing temperature trends from wind observations (Allen & Sherwood, 2008). The new observational estimates show greater warming than the earlier ones, and the new, larger set of estimates taken as a whole now bracket the trends predicted by the models (Thorne, 2008). This resolves a significant concern expressed in AR4 (Santer *et al.*, 2007).

WATER VAPOR, RAINFALL, AND THE HYDROLOGICAL CYCLE

New research and observations have resolved the question of whether a warming climate will lead to an atmosphere containing more water vapor, which would add to the greenhouse effect and enhance the warming. The answer is yes, this amplifying feedback has been detected: water vapor does become more plentiful in a warmer atmosphere (Dessler *et al.*, 2008). Satellite data show that atmospheric moisture content over the oceans has increased since 1998, with greenhouse emissions being the cause (Santer *et al.*, 2007).

No studies were cited in IPCC AR4 linking observed rainfall trends on a 50-year time scale to anthropogenic climate change. Now such trends can be linked. For example, Zhang *et al.* (2007) found that rainfall has reduced in the Northern Hemisphere subtropics but has increased in middle latitudes, and that this can be attributed to human-caused global warming. Models project that such trends will amplify as temperatures continue to rise.

Recent research has also found that rains become more intense in already-rainy areas as atmospheric water vapor content increases (Allan & Soden, 2008; Wentz *et al.*, 2007). Their conclusions strengthen those of earlier studies. However, recent changes have occurred even faster than predicted, raising the possibility that future changes could be more severe than predicted. This is a common theme from the recent science: uncertainties existing in AR4, once resolved, point to a more rapidly changing and more sensitive climate than we previously believed.

Q & A

Has Global Warming Recently Slowed Down or Paused?

No. There is no indication in the data of a significant slowdown or pause in the human-caused climatic warming trend. The observed global temperature changes are entirely consistent with the climatic warming trend of ~0.2 °C per decade predicted by IPCC, plus superimposed short-term variability (see Fig. 2.2). The latter has always been — and will always be — present in the climate system. Most of these short-term variations are due to internal oscillations like El Niño — Southern Oscillation, solar variability (predominantly the 11-year Schwabe cycle) and volcanic eruptions (which, like Pinatubo in 1991, can cause a cooling lasting a few years).

If one looks at periods of 10 years or shorter, such short-term variations can more than outweigh the anthropogenic global warming trend. For example, El Niño events typically come with global-mean temperature changes of up to 0.2 °C over a few years, and the solar cycle with warming or cooling of 0.1 °C over five years (Lean and Rind, 2008). However, neither El Niño, nor solar activity or volcanic eruptions make a significant contribution to longer-term climate trends. For good reason the IPCC has chosen 25 years as the shortest trend line they show in the global temperature records, and over this time period

---

Q & A   (*cont'd*)

the observed trend agrees very well with the expected anthropogenic warming.

Nevertheless global cooling has not occurred even since the exceptionally warm year 1998, contrary to claims promoted by lobby groups and picked up in some media. In all five available global temperature data sets the linear trend since 1998 is upward. In the NASA data it is 0.14 °C per decade, only slightly less than the 25-year trend of 0.18 °C per decade. The Hadley Center data most recently show smaller warming trends primarily due to the fact that this data set is not fully global but leaves out the Arctic, which has warmed particularly strongly in recent years.

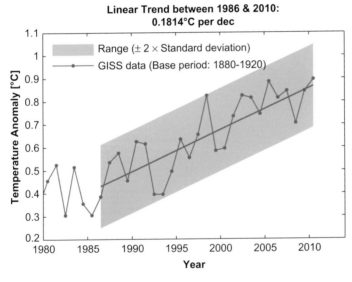

Linear Trend between 1986 & 2010:
0.1814°C per dec

FIGURE 2.2   Global temperature according to NASA GISS data since 1980. The blue line shows annual data, the straight line the 25-year linear trend, and the light blue shading the range of variability around this trend. Misunderstanding about warming trends can arise if only selected portions of the data are shown, e.g., 1998–2008, combined with the tendency to focus on extremes or end points (e.g., 2008 being cooler than 1998) rather than an objective trend calculation. Even the highly "cherry-picked" 11-year period starting with the warm 1998 and ending with the cold 2008 still shows a linear warming trend of 0.11 °C per decade.

(*Continued*)

Q & A (cont'd)

It is perhaps noteworthy that despite the extremely low brightness of the sun in recent years temperature records have been broken during this time (see NOAA, State of the Climate, 2009). For example, March 2008 saw a higher global land temperature than any March previously measured in the instrumental record. June and August 2009 saw higher land and ocean temperatures in the Southern Hemisphere than any previously recorded for those months. The global ocean surface temperatures in 2009 broke all previous records for three consecutive months: June, July, and August. The years 2007, 2008, and 2009 had the lowest summer Arctic sea ice cover ever recorded, and in 2008 for the first time in living memory the Northwest Passage and the Northeast Passage were simultaneously ice-free. This feat was repeated in 2009.

Q & A

Can Solar Activity or Other Natural Processes Explain Global Warming?

No. The incoming solar radiation has been almost constant over the past 50 years, apart from the well-known 11-year solar cycle (Fig. 2.3). In fact it has slightly decreased over this period. In addition, in recent years the brightness of the sun has reached an all-time low since the beginning of satellite measurements in the 1970s (Lockwood and Fröhlich, 2007, 2008). But this natural cooling effect was more than a factor of 10 smaller than the effect of increasing greenhouse gases, so it has not noticeably slowed down global warming. Also, winters are warming more rapidly than summers, and overnight minimum temperatures have warmed more rapidly than the daytime maxima — exactly the opposite of what would be the case if the sun were causing the warming.

Other natural factors, like volcanic eruptions or El Niño events, have only caused short-term temperature variations over time spans of a few years, but cannot explain any longer-term climatic trends (e.g., Lean and Rind, 2008).

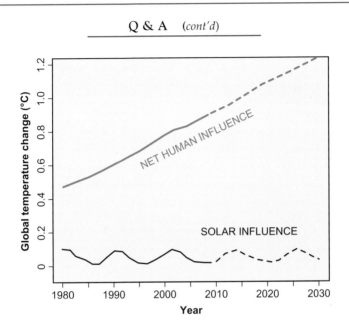

FIGURE 2.3   Time-series of solar irradiance alongside the net effect of green-house gas emissions (the latter relative to the year 1880; using Meehl et al., 2004) calculated in terms of total estimated impact on global air temperatures; observed from 1970–2008; and projected from 2009–2030 *(adapted from Lean and Rind, 2009)*

Extreme Events

KEY POINTS

- Increases in hot extremes and decreases in cold extremes have continued and are expected to amplify further.
- Anthropogenic climate change is expected to lead to further increases in precipitation extremes, both in heavy precipitation and in drought.
- Although future changes in tropical cyclone activity cannot yet be modeled, new analyses of observational data confirm that the intensity of tropical cyclones has increased in the past three decades in line with rising tropical ocean temperatures.

Many of the impacts of climate variations and climate change on society, the environment, and ecosystems arise through changes in the frequency or intensity of extreme weather and climate events. The IPCC Fourth Assessment Report (IPCC, 2007) concluded that many changes in extremes had been observed since the 1970s as part of the warming of the climate system. These included more frequent hot days, hot nights, and heat waves; fewer cold days, cold nights, and frosts; more frequent heavy precipitation events; more intense and longer droughts over wider areas; and an increase in intense tropical cyclone activity in the North Atlantic but no trend in total numbers of tropical cyclones.

TEMPERATURE EXTREMES

Recent studies have confirmed the observed trends of more hot extremes and fewer cold extremes and shown that these are consistent with the expected response to increasing greenhouse gases and anthropogenic aerosols at large spatial scales (Alexander & Arblaster, 2009; CCSP 2008a; Jones *et al.*, 2008; Meehl *et al.*, 2007a). However, at smaller scales, the effects of land-use change and variations of precipitation may be more important for changes in temperature extremes in some locations (Portmann *et al.*, 2009). Continued marked increases in hot extremes and decreases in cold extremes are expected in most areas across the globe due to further anthropogenic climate change (Alexander & Arblaster, 2009; CCSP, 2008a; Jones *et al.*, 2008; Kharin *et al.*, 2007; Meehl *et al.*, 2007a).

Precipitation Extremes and Drought

Post IPCC AR4 research has also found that rains become more intense in already-rainy areas as atmospheric water vapor content increases (Allan & Soden, 2008; Pall *et al.*, 2007; Wentz *et al.*, 2007). These conclusions strengthen those of earlier studies and are expected from considerations of atmospheric thermodynamics. However, recent changes have occurred faster than predicted by some climate models, raising the possibility that future changes will be more severe than predicted.

An example of recent increases in heavy precipitation is found in the United States, where the area with a much greater than normal proportion of days with extreme rainfall amounts has increased markedly (see Fig. 3.1). While these changes in precipitation extremes are consistent with

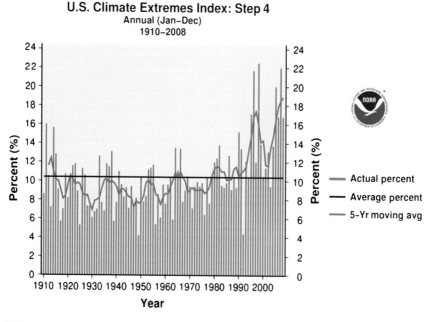

FIGURE 3.1   An increasing area of the United States is experiencing very heavy daily precipitation events. Annual values of the percentage area of the United States with a much greater than normal proportion of precipitation due to very heavy (equivalent to the highest tenth percentile) one-day precipitation amounts. From Gleason *et al.* (2008), updated by NOAA at /www.ncdc.noaa.gov/oa/climate/research/cei/cei.html.

the warming of the climate system, it has not been possible to clearly attribute them to anthropogenic climate change due to the very large variability of precipitation extremes (Alexander & Arblaster, 2009; CCSP, 2008a; Meehl *et al.*, 2007b).

In addition to the increases in heavy precipitation, there have also been observed increases in drought since the 1970s (Sheffield & Wood, 2008), consistent with the decreases in mean precipitation over land in some latitude bands that have been attributed to anthropogenic climate change (Zhang *et al.*, 2007).

The intensification of the global hydrological cycle with anthropogenic climate change is expected to lead to further increases in precipitation extremes, both in very heavy precipitation in wet areas and in drought in dry areas. While precise figures cannot yet be given, current studies suggest that heavy precipitation rates may increase by 5–10% per degree Celsius of warming, similar to the rate of increase of atmospheric water vapor.

Tropical Cyclones

The IPCC Fourth Assessment found a substantial upward trend in the severity of tropical cyclones (hurricanes and typhoons) since the mid-1970s, with a trend toward longer storm duration and greater storm intensity, strongly correlated with the rise in tropical sea surface temperatures. It concluded that a further increase in storm intensity is likely.

Several studies since the IPCC report have found more evidence for an increase in hurricane activity over the past decades. Hoyos *et al.* (2006) found a global increase in the number of hurricanes of the strongest categories 4 and 5, and they identified rising sea surface temperatures (SST) as the leading cause. Warming tropical SST has also been linked to increasingly intense tropical cyclone activity – and an increasing number of tropical cyclones – in the case of certain basins such as the North Atlantic (Emanuel *et al.*, 2008; Mann & Emanuel, 2006; Mann *et al.*, 2009a). Scientific debate about data quality has continued, especially on the question of how many tropical cyclones may have gone undetected before satellites provided a global coverage of observations. Mann *et al.*(2007) concluded that such an undercount bias would not be large enough to question the recent rise in hurricane activity and its close connection to sea surface warming. A complete reanalysis of satellite data since 1980 (Elsner *et al.*, 2008) confirms a global increase of the number of category 4 and 5 (i.e., the strongest) tropical cyclones: they found a 1 °C global warming corresponding to a 30% increase in these storms. While evidence has thus firmed up considerably that recent warming has been associated with stronger tropical cyclones, modeling studies (e.g., Emanuel *et al.*,

2008; Knutson *et al.*, 2008; Vecchi *et al.*, 2008) have shown that we have as yet no robust capacity to project future changes in tropical cyclone activity.

Other Severe Weather Events

The IPCC Fourth Assessment concluded that there were insufficient studies available to make an assessment of observed changes in small-scale severe weather events or of expected future changes in such events. However, recent research has shown an increased frequency of severe thunderstorms in some regions, particularly the tropics and south-eastern United States, which is expected due to future anthropogenic climate change (Aumann *et al.*, 2008; Marsh *et al.*, 2009; Trapp *et al.*, 2007, 2009). In addition, there have been recent increases in the frequency and intensity of wildfires in many regions with Mediterranean climates (e.g., Spain, Greece, southern California, south-east Australia) and further marked increases are expected due to anthropogenic climate change (Pitman *et al.*, 2007; Westerling *et al.*, 2006).

4

Land Surface

KEY POINTS

- Land cover change, particularly deforestation, can have a major impact on regional climate, but at the global scale its biggest impact comes from the CO_2 released in the process.
- Observations through the 2005 drought in Amazonia suggest that the tropical forests could become a strong carbon source if rainfall declines in the future.
- Carbon dioxide changes during the Little Ice Age indicate that warming may in turn lead to carbon release from land surfaces, a feedback that could amplify 21st century climate change.
- Avoiding tropical deforestation could prevent up to 20% of human-induced CO_2 emissions and help to maintain biodiversity.

HOW DOES LAND-USE CHANGE AFFECT CLIMATE?

Earth's climate is strongly affected by the nature of the land surface, including the vegetation and soil type and the amount of water stored on the land as soil moisture, snow, and groundwater. Vegetation and soils affect the surface albedo, which determines the amount of sunlight absorbed by the land. The land surface also affects the partitioning of rainfall into evapotranspiration (which cools the surface and moistens the atmosphere) and runoff (which provides much of our freshwater). This partitioning can affect local convection and therefore rainfall. Changes in land-use associated with the spread of agriculture, urbanization, and deforestation can alter these mechanisms. Land-use change can also change the surface roughness, and affect emissions of trace gases and some volatile organic compounds such as isoprene. Despite the key role of land cover change at regional scales, climate model projections from IPCC AR4 excluded anthropogenic land cover change.

There has been significant progress on modeling the role of land cover change since the IPCC AR4 (Piekle Sr *et al.*, 2007), with the first systematic study demonstrating that large-scale land cover change directly and significantly affects regional climate (Pitman *et al.*, 2009). This has important implications for understanding future climate change; climate models need to simulate land cover change to capture regional changes in regions of intense land cover change. However, failing to account for land cover change has probably not affected global-scale projections (Pitman *et al.*, 2009), noting that emissions from land cover change are included in projections.

Land cover change also affects climate change by releasing CO_2 to the atmosphere and thus modifying the land carbon sink (Bondeau *et al.*, 2007; Fargione *et al.*, 2008). The most obvious example of this is tropical deforestation which contributes about a fifth of global CO_2 emissions and also influences the land-to-atmosphere fluxes of water and energy (Bala *et al.*, 2007). Avoiding deforestation therefore eliminates a significant fraction of anthropogenic CO_2 emissions, and maintains areas like the Amazon rainforest which supports high biodiversity *and* plays a critically important role in the climate system (Malhi *et al.*, 2008).

CLIMATE CHANGE AND THE AMAZON RAINFOREST

The distribution and function of vegetation depends critically on the patterns of temperature and rainfall across the globe. Climate change therefore has the potential to significantly alter land cover even in the absence of land-use change. A key area of concern has been to preserve the remaining intact Amazonian rainforest which is susceptible to "dieback" in some climate models due to the combined effects of increasing greenhouse gases and reducing particulate or "aerosol" pollution in the Northern Hemisphere (Cox *et al.*, 2008). However, these projections are very dependent on uncertain aspects of regional climate change, most notably the sign and magnitude of rainfall change in Amazonia in the 21st century (Malhi *et al.*, 2008, 2009).

There have also been some doubts raised as to whether the Amazonian rainforest is as sensitive to rainfall reductions as large-scale models suggest. The drought in Western Amazonia in 2005 provided a test of this hypothesis using long-term monitoring of tree growth in the region (Phillips *et al.*, 2009), and a massive carbon source was detected in the region in 2005 against the backdrop of a significant carbon sink that occurred decades before. The forests of Amazonia are therefore sensitive to "2005-like" droughts and these are expected to become more common in the 21st century (Cox *et al.*, 2008).

A similar story emerges from the analysis of satellite and CO_2 flux measurements during the European drought of 2003 (Reichstein *et al.*,

2007). The IPCC AR4 tentatively suggested a link between global warming and the 2003 drought, and this analysis showed that the drought had an enormous impact on the health and functioning of both natural and managed landscapes in the region.

HOW LARGE ARE FEEDBACKS LINKING LAND SURFACE AND CLIMATE?

The response of the land surface to climatic anomalies feeds back on the climate by changing the fluxes of energy, water, and CO_2 between the land and the atmosphere. For example, it seems likely that changes in the state of the land surface, which in turn changed the energy and water fluxes to the atmosphere, played an important part in the severity and length of the 2003 European drought (Fischer et al., 2007). In some regions, such as the Sahel, land–atmosphere coupling may be strong enough to support two alternative climate-vegetation states: one wet and vegetated, the other dry and desert-like. There may be other "hot-spot" regions where the land–atmosphere coupling significantly controls the regional climate; indeed it appears that the land is a strong control on climate in many semi-arid and Mediterranean-like regions.

However, the strongest feedbacks on global climate in the 21st century are likely to be due to changes in the land carbon sink. The climate-carbon cycle models reported in the IPCC AR4 (Friedlingstein et al., 2006) reproduced the historical land carbon sink predominantly through "CO_2 fertilization." There is evidence of CO_2 fertilization being limited in nitrogen-limited ecosystems (Hyvönen et al., 2007), but the first generation coupled climate-carbon models did not include nutrient cycling.

The IPCC AR4 climate-carbon cycle models also represented a counteracting tendency for CO_2 to be released more quickly from the soils as the climate warms, and as a result these models predicted a reducing efficiency of the land carbon sink under global warming. There is some suggestion of a slow-down of natural carbon sinks in the recent observational record (Canadell et al., 2007), and strong amplifying land carbon-climate feedback also seems to be consistent with records of the Little Ice Age period (Cox & Jones, 2008).

DOES THE LAND SURFACE CARE ABOUT THE CAUSES OF CLIMATE CHANGE?

Yes. Vegetation is affected differently by different atmospheric pollutants, and this means that the effects of changes in atmospheric composition cannot be understood purely in terms of their impact on global warming.

CO_2 increases affect the land through climate change, but also directly through CO_2-fertilization of photosynthesis, and "CO_2-induced stomatal closure" which tends to increase plant's water-use efficiency. Observational studies have shown a direct impact of CO_2 on the stomatal pores of plants, which regulate the fluxes of water vapor and CO_2 at the leaf surface. In a higher CO_2 environment, stomata reduce their openings since they are able to take up CO_2 more efficiently. By transpiring less, plants increase their water-use efficiency, which consequently affects the surface energy and water balance. If transpiration is suppressed via higher CO_2, the lower evaporative cooling may also lead to higher temperatures (Cruz et al., 2010). There is also the potential for significant positive impacts on freshwater resources, but this is still an area of active debate (Betts et al., 2007; Gedney et al., 2006; Piao et al., 2007).

By contrast, increases in near-surface ozone have strong negative impacts on vegetation by damaging leaves and their photosynthetic capacity. As a result, historical increases in near-surface ozone have probably suppressed land carbon uptake and therefore increased the rate of growth of CO_2 in the 20th century. Sitch et al. (2007) estimate that this indirect forcing of climate change almost doubles the contribution that near-surface ozone made to 20th century climate change.

Atmospheric aerosol pollution also has a direct impact on plant physiology by changing the quantity and nature of the sunlight reaching the land surface. Increasing aerosol loadings from around 1950 to 1980, associated predominantly with the burning of sulphurous coal, reduced the amount of sunlight at the surface, which has been coined "global dimming" (Wild et al., 2007). Since plants need sunlight for photosynthesis, we might have expected to see a slow-down of the land carbon sink during the global dimming period, but we didn't. Mercado et al. (2009) offer an explanation for this based on the fact that plants are more light-efficient if the sunlight is "diffuse." Aerosol pollution would certainly have scattered the sunlight, making it more diffuse, as well as reducing the overall quantity of sunlight reaching the surface. It seems that "diffuse radiation fertilization" won this battle, enhancing the global land carbon sink by about a quarter from 1960 to 2000 (Mercado et al., 2009). This implies that the land carbon sink will decline if we reduce the amount of potentially harmful particulates in the air.

These recent studies since IPCC AR4 argue strongly for metrics to compare different atmospheric pollutants that go beyond radiative forcing and global warming, to impacts on the vital ecosystem services related to the availability of food and water.

CHAPTER

5

Permafrost and Hydrates

KEY POINTS

- New insights into the Northern Hemisphere permafrost (permanently frozen ground) suggest a large potential source of CO_2 and CH_4 that would amplify atmospheric concentrations if released.

- A recent increase in global methane levels cannot yet be attributed to permafrost degradation.

- A separate and significant source of methane exists as hydrates beneath the deep ocean floor and in permafrost. It has recently been concluded that release of this type of methane is very unlikely to occur this century.

As noted in the IPCC AR4 and more recent studies, the southern boundary of the discontinuous permafrost zone has shifted northward over North America in recent decades. Rapid degradation and upward movement of the permafrost lower limit have continued on the Tibetan plateau (Cui & Graf, 2009; Jin et al., 2008). In addition, observations in Europe (Åkerman & Johansson, 2008; Harris et al., 2009) have noted permafrost thawing and a substantial increase in the depth of the overlying active layer exposed to an annual freeze/thaw cycle, especially in Sweden.

As permafrost melts and the depth of the active layer deepens, more organic material can potentially start to decay. If the surface is covered with water, methane-producing bacteria break down the organic matter. But these bacteria cannot survive in the presence of oxygen. Instead, if the thawed soils are exposed to air, carbon dioxide-producing bacteria are involved in the decay process. Either case is an amplifying feedback to global warming. In fact, the magnitude of the feedback represents an important unknown in the science of global warming; this feedback has not been accounted for in any of the IPCC projections. The total amount of carbon stored in permafrost has been estimated to be around 1672 Gt (1 Gt = 10^9 tons), of which ~277 Gt are contained in peatlands

(Schuur *et al.*, 2008; Tarnocai, *et al.*, 2009). This represents about twice the amount of carbon contained in the atmosphere. A recent analysis by Dorrepaal *et al.* (2009) has found strong direct observational evidence for an acceleration of carbon emissions in association with climate warming from a peat bog overlying permafrost at a site in northern Sweden. It is still uncertain whether recent observations of increasing atmospheric methane concentration (Rigby *et al.*, 2008), after nearly a decade of stable levels, are caused by enhanced northern hemisphere production associated with surface warming.

Another amplifying feedback to warming that has recently been observed in high northern latitudes involves the microbial transformation of nitrogen trapped in soils to nitrous oxide. By measuring the nitrous oxide emissions from bare peat surfaces, Repo *et al.* (2009) inferred emissions per square meter of the same magnitude as those from croplands and tropical soils. They point out that as the Arctic warms, regions of bare exposed peat will increase, thereby amplifying total nitrous oxide emissions.

Between 500 and 10,000 Gt of carbon are thought to be stored under the sea floor in the form of methane hydrates (or clathrates), a crystalline structure of methane gas and water molecules (Brook *et al.*, 2008). Another 7.5 to 400 Gt of carbon are stored in the form of methane hydrates trapped in permafrost (Brook *et al.*, 2008). Some have argued that anthropogenic warming could raise the possibility of a catastrophic release of methane from hydrates to the atmosphere. In a recent assessment by the US Climate Change program (CCSP, 2008b), it was deemed to be very unlikely that such a release would occur this century, although the same assessment deemed it to be very likely that methane sources from hydrate and wetland emissions would increase as the climate warmed. This is supported by a recent analysis which found that the observed increase in atmospheric methane 11,600 years ago had a wetland, as opposed to hydrate, origin (Petrenko *et al.*, 2009)—as was also found in studies using earth models of intermediate complexity (Archer *et al.*, 2009; Fyke & Weaver, 2006).

Few studies with AR4-type climate models have been undertaken. One systematic study used the Community Climate System Model, version 3 (CCSM3), with explicit treatment of frozen soil processes. The simulated reduction in permafrost reached 40% by ~2030 irrespective of emission scenario (a reduction from ~10 million km^2 to 6 million km^2). By 2050, this reduces to 4 million km^2 (under B1 emissions) and 3.5 million km^2 (under A2 emissions). Permafrost declines to ~1 million km^2 by 2100 under A2. In each case, the simulations did not include additional feedbacks triggered by the collapse of permafrost including out-gassing of methane, a northward expansion of shrubs and forests, and the activation of the soil carbon pool. These would each further amplify warming.

6

Mountain Glaciers and Ice Caps

KEY POINTS

- There is widespread evidence of increased melting of mountain glaciers and ice caps since the mid-1990s.
- The contribution of mountain glaciers and ice caps to global sea level has increased from 0.8 mm per year in the 1990s to 1.2 mm per year today.
- The adjustment of mountain glaciers and ice caps to present climate alone is expected to raise sea level by ~18 cm. Under warming conditions they may contribute as much as ~55 cm by 2100.

Mountain glaciers and ice caps can potentially contribute a total of approximately 0.7 m to global sea level. They also provide a source of freshwater in many mountain regions worldwide. The IPCC AR4 assessed the contribution from worldwide shrinking mountain glaciers and ice caps to sea level rise at the beginning of the 21st century at about 0.8 mm per year (Kaser *et al.*, 2006; Lemke *et al.*, 2007). Since then, new estimates of the contribution from mountain glaciers and ice caps have been made using new data and by exploring new assessment methods.

These new assessments are shown in Fig. 6.1. They show mountain-glacier and ice-cap contributions to sea level rise that are generally slightly higher than those reported in IPCC AR4. They also extend from 1850 up to 2006. These new estimates show that the mass loss of mountain glaciers and ice caps has increased considerably since the beginning of the 1990s and now contributes about 1.2 mm per year to global sea level rise.

Mountain glaciers and ice caps are not in balance with the present climate. Recent estimates show that adjustment to that alone will cause a mass loss equivalent to ~18 cm sea level rise (Bahr *et al.*, 2009) within this century. Under ongoing changes consistent with current warming trends, a mass loss of up to ~55 cm sea level rise is expected by 2100 (Pfeffer *et al.*, 2008).

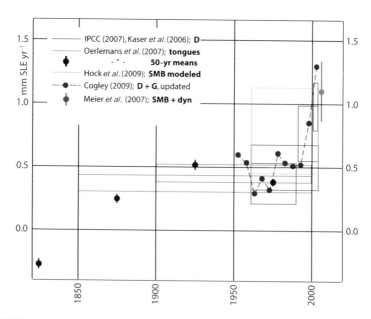

FIGURE 6.1   Estimates of the contribution of mountain glaciers and ice caps to global change in sea level equivalent (SLE), in millimeters SLE per year.

7

Ice Sheets of Greenland and Antarctica

KEY POINTS

- The surface area of the Greenland ice sheet which experiences summer melt has increased by 30% since 1979, consistent with warming air temperatures. Melt covered 36% of the ice sheet during the record season in 2007.

- The net loss of ice from the Greenland ice sheet has accelerated since the mid-1990s and is now contributing as much as 0.7 mm per year to sea level rise due to increased melting and accelerated ice flow.

- Antarctica is also losing ice mass at an increasing rate, mostly from the West Antarctic ice sheet due to increased ice flow. Antarctica is currently contributing to sea level rise at a rate nearly equal to Greenland.

Antarctica and Greenland maintain the largest ice reservoirs on land. If completely melted, the Antarctic ice sheet would raise global sea level by 52.8 m, while Greenland would add a further 6.6 m. Loss of only the most vulnerable parts of West Antarctica would still raise sea level by 3.3 m (Bamber *et al.*, 2009). IPCC AR4 concluded that net ice loss from the Greenland and Antarctic ice sheets together contributed to sea level rise over the period 1993 to 2003 at an average rate estimated at 0.4 mm per year. Since IPCC AR4, there have been a number of new studies observing and modeling ice sheet mass budget that have considerably enhanced our understanding of ice sheet vulnerabilities (Allison *et al.*, 2009). These assessments reinforce the conclusion that the ice sheets are contributing to present sea level rise, and show that the rate of loss from both Greenland and Antarctica has increased recently. Furthermore, recent observations have shown that changes in the rate of ice discharge into the sea can occur far more rapidly than previously suspected (e.g., Rignot, 2006).

GREENLAND

Figure 7.1 shows estimates of the mass budget of the Greenland ice sheet since 1960. In this representation, the horizontal dimension of the boxes shows the time period over which the estimate was made, and the vertical dimension shows the upper and lower limits of the estimate. The colors represent the different methods that were used: estimates derived from satellite or aircraft altimeter measurements of height change of the ice sheet surface are brown; estimates of mass loss from satellite gravity measurements are blue; and estimates derived from the balance between mass influx and discharge are red.

The data in Fig. 7.1 indicate that net ice mass loss from Greenland has been increasing since at least the early 1990s, and that in the 21st century, the rate of loss has increased significantly. Multiple observational constraints and the use of several different techniques provide confidence that the rate of mass loss from the Greenland ice sheet has accelerated. Velicogna (2009) used GRACE satellite gravity data to show that the Greenland mass loss doubled over the period from April 2002 to February 2009.

Near-coastal surface melt and run-off have increased significantly since 1960 in response to warming temperature, but total snow precipitation has also increased (Hanna *et al.*, 2008). The average Greenland surface temperature rose by more than 1.5 °C over the period 2000 to 2006

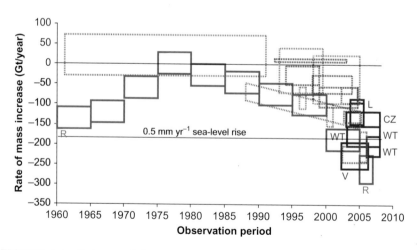

FIGURE 7.1   Estimates of the net mass budget of the Greenland ice sheet since 1960. A negative mass budget indicates ice loss and sea level rise. Dotted boxes represent estimates used by IPCC AR4 (IPCC, 2007). The solid boxes are post-AR4 assessments (R = Rignot et al., 2008a; VW = Velicogna & Wahr, 2006; L = Luthcke et al., 2006; WT = Wouters et al., 2008; CZ = Cazenave et al., 2009; V = Velicogna, 2009).

and mass loss estimated from GRACE gravity data occurred within 15 days of the initiation of surface melt, suggesting that the water drains rapidly from the ice sheet (Hall *et al.*, 2008). Passive microwave satellite measurements of the area of the Greenland ice sheet subject to surface melt indicate that the melt area has been increasing since 1979 (Steffen *et al.*, 2008; Fig. 7.2). There is a good correlation between total melt area extent and the number of melt days with total volume of run off, which has also increased.

The pattern of ice sheet change in Greenland is one of near-coastal thinning, primarily in the south along fast-moving outlet glaciers. Accelerated flow and discharge from some major outlet glaciers (also called dynamic thinning) is responsible for much of the loss (Howat *et al.*, 2007; Rignot & Kanagaratnam, 2006). In southeast Greenland many smaller drainage basins, especially the catchments of marine-terminating outlet glaciers, are also contributing to ice loss (Howat *et al.*, 2008). Pritchard *et al.* (2009) used high resolution satellite laser altimetry to show

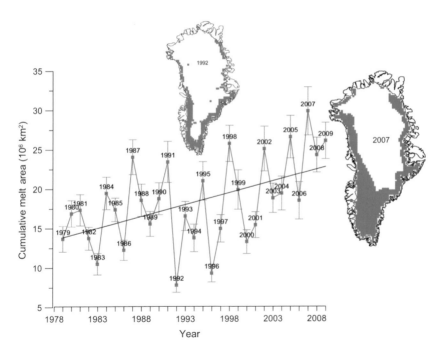

FIGURE 7.2   The total melt area of the Greenland ice sheet increased by 30% between 1979 and 2009 based on passive microwave satellite data, with the most extreme melt in 2007. In general, 33–55% of the total mass loss from the Greenland ice sheet is caused by surface melt and runoff. For 2007, the area experiencing melt was around 36% of the total ice sheet area. The low melt year in 1992 was caused by the volcanic aerosols from Mt. Pinatubo causing a short-lived global cooling (updated from Steffen et al., 2008).

that dynamic thinning of fast-flowing coastal glaciers is now widespread at all latitudes in Greenland. Greenland glaciers flowing faster than 100 m per year thinned by an average of 0.84 m per year between 2003 and 2007.

ANTARCTICA

New estimates of the mass budget of the Antarctic ice sheet are shown in Fig. 7.3. Comprehensive estimates for Antarctica are only available since the early 1990s. Several new studies using the GRACE satellite gravity data (blue boxes in Fig. 7.3) all show net loss from the Antarctic since 2003 with a pattern of near balance for East Antarctica, and greater mass loss from West Antarctica and the Antarctic Peninsula (e.g., Cazenave *et al.*, 2009; Chen *et al.*, 2006). The GRACE assessment of Velicogna (2009) indicates that, like Greenland, the rate of mass loss from the Antarctic ice sheet is accelerating, increasing from 104 Gt per year for 2002–2006 to 246 Gt per year for 2006–2009 (the equivalent of almost 0.7 mm per year of sea level rise). Gravity and altimeter observations require correction for uplift of the Earth's crust under the ice sheets (glacial isostatic adjustment): this is poorly known for Antarctica.

The largest losses occurred in the West Antarctic basins draining into the Bellingshausen and Amundsen seas. Satellite glacier velocity estimates from 1974 imagery show that the outlet glaciers of the Pine Island Bay region have accelerated since then, changing a region of the ice sheet that was in near balance to one of considerable loss (Rignot, 2008). Rignot

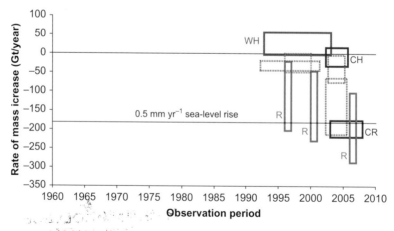

FIGURE 7.3 Estimates of the net mass budget of the Antarctic ice sheet since 1992. Dotted boxes represent estimates used by IPCC AR4 (IPCC, 2007). The solid boxes are more recent estimates (CH = Chen *et al.*, 2006; WH = Wingham *et al.*, 2006; R = Rignot *et al.*, 2008b).

et al. (2008b) show that the ice discharge in this region further increased between 1996 and 2006, increasing the net mass loss over the period by 59%, and Pritchard *et al.* (2009) show from laser altimetry that dynamic thinning in some parts of the Amundsen Sea embayment has exceeded 9 m per year. The recent acceleration of ice streams in West Antarctica explains much of the Antarctic mass loss, but narrow fast-moving ice streams in East Antarctica are also contributing to the loss (Pritchard *et al.*, 2009).

The Antarctic Peninsula region has experienced much greater warming than the continent as a whole. This has led to widespread retreat (Cook *et al.*, 2005) and acceleration (Pritchard & Vaughan, 2007) of the tidewater glaciers in that region.

THE RISK OF ICE SHEET COLLAPSE

The largest unknown in the projections of sea level rise over the next century is the potential for rapid dynamic collapse of ice sheets. The most significant factor in accelerated ice discharge in both Greenland and Antarctica over the last decade has been the un-grounding of glacier fronts from their bed, mostly due to submarine ice melting. Changes to basal lubrication by melt water, including surface melt draining through moulins (vertical conduits) to the bottom of the ice sheet, may also affect the ice sheet dynamics in ways that are not fully understood. The major dynamic ice sheet uncertainties are largely one-sided: they can lead to a faster rate of sea level rise, but are unlikely to significantly slow the rate of rise. Although it is unlikely that the total sea level rise by 2100 will be as high as 2 m (Pfeffer *et al.*, 2008), the probable upper limit of a contribution from the ice sheets remains uncertain.

Ice Shelves

KEY POINTS

- Ice shelves connect continental ice sheets to the ocean. Destabilization of ice shelves along the Antarctic Peninsula has been widespread with seven collapses over the past 20 years.

- Signs of ice shelf weakening have been observed elsewhere than in the Antarctic Peninsula, e.g. in the Bellingshausen and Amundsen seas, indicating a more widespread influence of atmospheric and oceanic warming than previously thought.

- There is a strong influence of ocean warming on ice sheet stability and mass balance via the melting of ice shelves.

Ice shelves are floating sheets of ice of considerable thickness that are attached to the coast. They are mostly composed of ice that has flowed from the interior ice sheet, or that has been deposited as local snowfall. They can be found around 45% of the Antarctic coast, in a few bays off the north coast of Ellesmere Island near Greenland, and in a few fiords along the northern Greenland coast (where they are termed ice tongues). Over the last few years, the six remaining ice shelves (Serson, Petersen, Milne, Ayles, Ward Hunt, and Markham) off Ellesmere Island have either collapsed entirely (Ayles on August 13, 2005, and Markham during the first week of August, 2008) or undergone significant disintegration.

Along the coast of Greenland, the seaward extent of the outlet glacier *Jakobshavn Isbrae* provides a striking example of a floating ice tongue in retreat (Fig. 8.1). Holland *et al.* (2008) suggest that the observed recent acceleration (Rignot & Kanagaratnam, 2006) of Jakobshavn Isbrae may be attributed to thinning from the arrival of warm waters in the region.

Destabilization of floating ice shelves has been widespread along the Antarctic Peninsula with seven collapsing in the last 20 years. Warming along the Peninsula has been dramatic, and on the western side has been

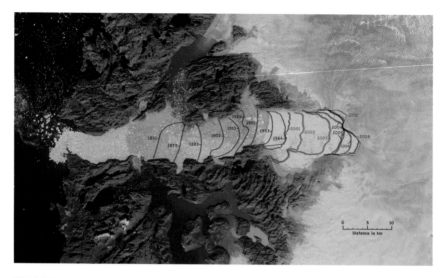

FIGURE 8.1   Photograph of the floating ice tongue representing the seaward extent of
Jakobshavn Isbrae. The Landsat image shown in the background is a false-color image of
data collected on July 29, 2009. Changes in the position of the calving front from 1851 to 2010
are indicated. Credit: NASA/Goddard Space Flight Center Scientific Visualization Studio,
http://svs.gsfc.nasa.gov/vis/a000000/a003800/a003806/

substantially above the global average. Most recently, in March 2009,
more than 400 km^2 collapsed off the Wilkins Ice Shelf on the western side
of the Antarctic Peninsula. A number of mechanisms are thought to play
important roles in destabilizing floating Antarctic ice shelves. These
include: surface warming leading to the creation of melt ponds and
subsequent fracturing of existing crevasses (van den Broeke, 2005);
subsurface ice shelf melting from warming ocean waters (Rignot et al.,
2008); and internal ice shelf stresses (Braun & Humbert, 2009). While the
collapse of a floating ice shelf does not itself raise sea level, its collapse is
followed by rapid acceleration of glacier outflow — which does raise sea
level — due to the removal of the ice shelf buttressing effect (e.g., Rignot
et al., 2004; Scambos et al., 2004).

There is evidence for the melting of ice shelves in the Amundsen Sea,
with impacts on the flow speed of glaciers draining this part of West
Antarctica. A recent modeling study has suggested that the West
Antarctic ice sheet would begin to collapse when ocean temperatures
in the vicinity of any one of these ice shelves that surround it warm
by about 5 °C (Pollard & DeConto, 2009). There is also evidence that
these changes are not limited to West Antarctica and may also affect
the coastline of East Antarctica, for example in Wilkes Land (Pritchard
et al., 2009; Shepherd & Wingham, 2007). The widespread thinning

and acceleration of glaciers along the Antarctic coast may indicate a significant impact of oceanic changes on glacier dynamics, a factor that has received little attention in past IPCC reports due to the lack of observational data on ice—ocean interactions and how climate change might influence coastal ocean waters.

9

Sea Ice

ARCTIC SEA ICE

Perhaps the most stunning observational change since the IPCC AR4 has been the shattering of the previous Arctic summer minimum sea ice extent record — something not predicted by climate models. Averaged over the five-day period leading up to September 16, 2007, the total extent of sea ice in the Arctic was reduced to an area of only 4.1 million km^2 (see Fig. 9.1), surpassing the previous minimum set in 2005 by 1.2 million km^2 (about the same size as France, Spain, Portugal, Belgium, and Netherlands combined). The median September minimum sea ice extent since observations with the current generation of multi-frequency passive microwave sensors commenced in 1979–2000 was 6.7 million km^2. Compared to the median, the 2007 record involved melting 2.6 million km^2 more ice (~40% of the median).

The September Arctic sea ice extent over the last several decades has decreased at a rate of 11.1 ± 3.3% per decade (NSIDC, 2009). This dramatic

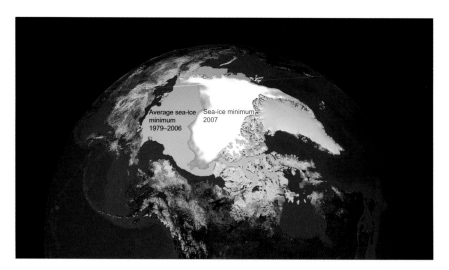

FIGURE 9.1 Arctic sea ice extent over the five days leading up to and including September 16, 2007, compared to the average sea ice minimum extent for the period 1979–2006. *Source: NASA/Goddard Space Flight Center Scientific Visualization Studio.*

retreat has been much faster than that simulated by any of the climate models assessed in the IPCC AR4 (Fig. 9.2). This is likely due to a combination of several model deficiencies, including: (1) incomplete representation of ice albedo physics, including the treatment of melt

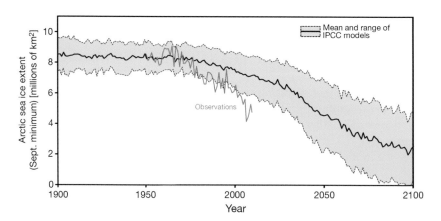

FIGURE 9.2 Observed (red line) and modeled September Arctic sea ice extent in millions of square kilometers. The solid black line gives the ensemble mean of the 13 IPCC AR4 models while the dashed black lines represent their range. From Stroeve *et al.* (2007) updated to include data for 2008 and 2009. The 2010 minimum has recently been calculated at 4.6 million km^2, the third lowest year on record, and still well below the IPCC worst case scenario.

ponds (e.g., Pedersen *et al.*, 2009), and the deposition of black carbon (e.g., Flanner *et al.*, 2007; Ramanathan & Carmichael, 2008); and (2) incomplete representation of the physics of vertical and horizontal mixing in the ocean (e.g., Arzel *et al.*, 2006). Winter Arctic sea ice extent has also decreased since 1979, but at a slower rate than in summer. The February extent has decreased at a rate of 2.9 ± 0.8% per decade (NSIDC, 2009).

The thickness of Arctic sea ice has also been on a steady decline over the last several decades. For example, Lindsay *et al.* (2009) estimated that the September sea ice thickness has been decreasing at a rate of 57 cm per decade since 1987. Similar decreases in sea ice thickness have been detected in winter. For example, within the area covered by submarine sonar measurements, Kwok and Rothrock (2009) show that the overall mean winter thickness of 3.64 m in 1980 decreased to only 1.89 m by 2008—a net decrease of 1.75 m, or 48%. By the end of February 2009, less than 10% of Arctic sea ice was more than two years old, down from the historic values of 30%.

WHEN WILL THE ARCTIC OCEAN BE ICE-FREE?

Due to the existence of natural variability within the climate system, it is not possible to predict the precise year that the Arctic Ocean will become seasonally ice-free. Nevertheless, the warming commitment associated with existing atmospheric greenhouse gas levels very likely means that a summer ice-free Arctic is inevitable. Evidence is also emerging to suggest that the transition to an ice-free summer in the Arctic might be expected to occur abruptly, rather than slowly (Holland *et al.*, 2006), because of amplifying feedbacks inherent within the Arctic climate system. In fact, in one of the simulations of the NCAR Climate System Model version 3 (CCSM3) discussed in Holland *et al.* (2006), the Arctic summer became nearly ice-free by 2040. As noted by Lawrence *et al.* (2008), an abrupt reduction in Arctic summer sea ice extent also triggers rapid warming on land and subsequent permafrost degradation.

ANTARCTIC SEA ICE

Unlike the Arctic, Antarctic sea ice extent changes have been more subtle, with a net annual-mean area increase of ~1% per decade over the period 1979–2006 (Cavalieri & Parkinson, 2008; Comiso & Nishio, 2008). However, there have been large regional changes in Antarctic sea ice distribution: for example, the Weddell and Ross Sea areas have shown increased extent linked to changes in large-scale atmospheric circulation, while the western Antarctic Peninsula region and the coast of West

Antarctica (Amundsen and Bellingshausen seas) show a significant decline consistent with more northerly winds and surface warming observed there (Lefebvre et al., 2004; Steig et al., 2009; Turner et al., 2009). These regional changes are linked to a major change in the seasonality of the ice; that is, its duration and the timing of the annual advance and retreat (Stammerjohn et al., 2008).

Since Antarctica is a land mass surrounded by the vast Southern Ocean, whereas the Arctic is a small ocean surrounded by vast amounts of land, and as oceans respond less rapidly than land to warming because of their thermal stability, one would expect, and indeed climate models show, a delayed warming response around Antarctica. In addition, Turner et al. (2009) note that stratospheric ozone depletion arising from the anthropogenic release of chlorofluorocarbons (CFCs) has led to the strengthening of surface winds around Antarctica during December to February (summer). They argue that these strengthened winds are in fact the primary cause for the slight positive trend in Antarctic sea ice extent observed over the last three decades. However, as CFCs are regulated under the Montreal Protocol and have declining atmospheric concentrations, the ozone hole over Antarctica is expected to recover and hence one anticipates an acceleration of sea ice melt in the Southern Hemisphere in the decades ahead.

There are few data available on the thickness distribution of Antarctic pack ice, and no information on any changes in the thickness of Antarctic sea ice.

Q & A

Isn't Antarctica Cooling and Antarctic Sea Ice Increasing?

Antarctica is not cooling: it has warmed overall over at least the past 50 years. Although the weather station at the South Pole shows cooling over this period, this single weather station is not representative. For example, there is a warming trend at Vostok, the only other long-term monitoring station in the interior of the continent. Several independent analyses (Chapman & Walsh, 2007; Goosse et al., 2009; Monaghan et al., 2008; Steig et al., 2009) show that on average, Antarctica has warmed by about 0.5 °C since measurements began in the 1957 International Geophysical Year, with particularly rapid warming around the Antarctic Peninsular region and over the West Antarctic ice sheet (Fig. 9.3 shows the mean trend from 1957–2006). Furthermore, there is direct evidence from borehole measurements that warming in West Antarctica began no later than the 1930s (Barrett et al., 2009).

Q & A   *(cont'd)*

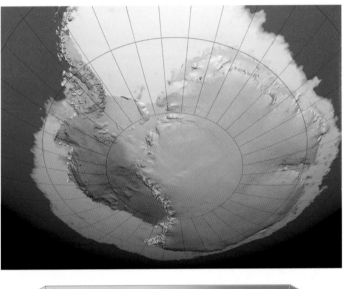

FIGURE 9.3 Annual mean air temperature trend in °C/decade during 1957–2006 from Steig *et al.* (2009). Adapted by permission from Macmillan Publishers Ltd: *Nature* Vol. 457, Steig et al, "Warming of the Antarctic ice-sheet surface since the 1957 International Geophysical Year" p. 459-462, Fig 3, copyright 2008.

Since the development of the Antarctic ozone hole in the late 1970s, there has been a strengthening of the circumpolar winds around Antarctica, which tends to reduce the amount of warmer air reaching the interior of the continent. The stronger winds are due to cooling in the upper atmosphere, which are in turn a result of ozone depletion caused by chlorofluorocarbons. As a consequence, much of East Antarctica has cooled in summer and fall seasons since the late 1970s. Ironically, human emissions of CFCs are thus helping to partly offset interior Antarctic warming, analogous to the global dimming due to sulphate aerosols. As the ozone hole gradually repairs over the coming century, the cooling offset is likely to diminish.

(Continued)

Q & A   (*cont'd*)

The factors that govern sea ice extent around Antarctica are very different from the situation in the Arctic, because Antarctica is a continent sitting on the pole and surrounded by water, just the opposite of the Arctic geography. The extent of sea ice around Antarctica is strongly determined by the circumpolar winds which spread the ice out from the continent, and by the position of the polar front where the ice encounters warmer ocean waters. Sea ice cover in Antarctica shows a slight (not statistically significant) upward trend, consistent with the increase in circumpolar winds mentioned above. In West Antarctica, where the temperature increases are the greatest, sea ice has declined at a statistically significant rate since at least the 1970s.

10

The Oceans

KEY POINTS

- Estimates of ocean heat uptake have converged and are found to be 50% higher than previous calculations.

- Global ocean surface temperature reached the warmest ever recorded for each of June, July, and August 2009.

- Ocean acidification and ocean de-oxygenation have been identified as potentially devastating for large parts of the marine ecosystem.

Detection of how climate change is impacting the oceans has improved markedly since the IPCC AR4. Significant changes in temperature, salinity, and biogeochemical properties have been measured. These changes are consistent with the observed 50-year warming, rainfall, and CO_2 trends in the atmosphere. There have also been important new analyses of the trends in a broader range of properties since the IPCC AR4, including acidification and oxygen. This has improved our understanding of the changing state of the oceans and also identified new issues. Where new estimates of ocean change exist since IPCC AR4, they tend to be larger and also more consistent with projections of climate change (e.g., global heat content and sea level).

OCEAN WARMING

There has been a long-term sustained warming trend in ocean surface temperatures over the past 50 years (Fig. 10.1). Surface ocean warming from the decade 1991–2000 to the decade 2001–2010 was approximately 0.1 degrees C (according to the NOAA sea surface temperature), higher than any previous decade-to-decade rise.

Increases in oceanic heat content in the upper ocean (0–700m) between 1963 and 2003 have been found to be 50% higher than previous estimates

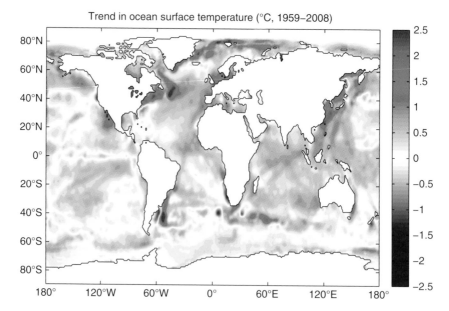

FIGURE 10.1   Long-term 50-year change in sea surface temperature (SST) during 1959–2008 calculated by fitting a linear trend to 50 years of monthly SST data at each grid point. The SST fields are from the Hadley Centre data set as described by Rayner *et al.* (2006).

(Bindoff *et al.*, 2007; Domingues *et al.*, 2008). The higher estimates of heat content change are now consistent with observations of sea level rise over the last 50 years, resolving a long-standing scientific problem in understanding the contribution of thermal expansion to sea level (Domingues *et al.*, 2008). Observations also show deep-ocean warming that is much more widespread in the Atlantic and Southern oceans (Johnson *et al.*, 2008a, 2008b) than previously appreciated.

SALINITY AND THE HYDROLOGICAL CYCLE

More comprehensive analyses of ocean salinity show a freshening of high latitudes, while regions of excess evaporation over precipitation have become saltier. The salinity changes are consistent with a strengthening of the hydrological cycle. The patterns of salinity change are also consistent with regional circulation and inter-basin exchanges. We now have increased evidence that the long-term trends in patterns of rainfall over the global ocean, as reflected in salinity, can be attributed to climate change (Stott *et al.*, 2008).

CLIMATE CHANGE AND OCEAN CIRCULATION

Surprising salinity changes in Antarctic bottom waters provide additional evidence of increased melt from the ice sheets and ice shelves (Rintoul, 2007). The Arctic shows strong evidence for increased precipitation and river run-off. Intermediate layers in the Arctic Ocean have warmed notably (Polyakov et al., 2004). Consistent with current model results, observations are yet to detect any indication of a sustained change in the North Atlantic Ocean circulation (e.g., Hansen & Østerhus, 2007).

Regional climate change is often organized and expressed around the main patterns of variation such as the North Atlantic Oscillation, El Niño, and the Southern Annular Mode. These patterns themselves may be affected by greenhouse gases, leading to either larger fluctuations, or a preferred state in coming decades (e.g., a trend toward a different type of El Niño event) (Latif & Keenlyside, 2009; Yeh et al., 2009). Currently the influence of regional climate modes on ocean circulation is larger than the underlying trends attributable to anthropogenic climate change.

The stability of the North Atlantic Ocean circulation is vitally important for North American and European climate. For example, a slowdown of these ocean currents could lead to a more rapid rise of regional sea level along the northeast US coast (Yin et al., 2009). The IPCC AR4 concluded that there is greater than 90% probability of a slowdown of this ocean current system, and less than 10% risk of a "large abrupt transition" by the year 2100. As noted in the Synthesis and Assessment Project 3.4 of the US Climate Change Science Program (Delworth et al., 2008), no comprehensive climate model projects such a transition within this century. However, given uncertainty in our ability to model nonlinear threshold behavior and the recent suggestion that models may be too stable (Hofmann & Rahmstorf, 2009), we cannot completely exclude the possibility of such an abrupt transition.

OCEAN ACIDIFICATION, CARBON UPTAKE, AND OCEAN DE-OXYGENATION

The CO_2 content of the oceans increased by 118 ± 19 Gt (1 Gt = 10^9 tons) between the end of the pre-industrial period (about 1750) and 1994, and continues to increase by about 2 Gt each year (Sabine et al., 2004). The increase in ocean CO_2 has caused a direct decrease in surface ocean pH by an average of 0.1 units since 1750 and an increase in acidity by more than 30% (McNeil & Matear, 2007; Orr et al., 2005; Riebesell et al., 2009). Calcifying organisms and reefs have been shown to be particularly vulnerable to high CO_2, low pH waters (Fabry et al., 2008).

New in-situ evidence shows a tight dependence between calcification and atmospheric CO_2, with smaller shells evident during higher CO_2 conditions over the past 50,000 years (Moy *et al.*, 2009). Furthermore, due to preexisting conditions, the polar oceans of the Arctic and Southern oceans have been shown to start dissolving certain shells once the atmospheric levels reach 450 ppm (~2030 under business-as-usual; McNeil & Matear, 2008; Orr *et al.*, 2009).

There is new evidence for a continuing decrease in dissolved oxygen concentrations in the global oceans (Oschlies *et al.*, 2008), and there is for the first time significant evidence that the large equatorial oxygen minimum zones are already expanding in a warmer ocean (Stramma *et al.*, 2008). Declining oxygen is a stress multiplier that causes respiratory issues for large predators (Rosa & Seibel, 2008) and significantly compromises the ability of marine organisms to cope with acidification (Brewer, 2009). Increasing areas of marine anoxia have profound impacts on the marine nitrogen cycle, with yet unknown global consequences (Lam *et al.*, 2009). A recent modeling study (Hofmann & Schellnhuber, 2009) points to the risk of a widespread expansion of regions lacking in oxygen in the upper ocean if increases in atmospheric CO_2 continue.

11

Global Sea Level

KEY POINTS

- Satellite measurements show sea level is rising at 3.4 mm per year since these records began in 1993. This is 80% faster than the best estimate of the IPCC Third Assessment Report for the same time period.

- Accounting for ice sheet mass loss, sea level rise until 2100 is likely to be at least twice as large as that presented by IPCC AR4, with an upper limit of ~2 m based on new ice sheet understanding.

Population densities in coastal regions and on islands are about three times higher than the global average. Currently 160 million people live less than 1 m above sea level. This allows even small sea level rise to have significant societal and economic impacts through coastal erosion, increased susceptibility to storm surges and resulting flooding, ground-water contamination by salt intrusion, loss of coastal wetlands, and other issues.

Since 1870, global sea level has risen by about 20 cm (IPCC AR4). Since 1993, sea level has been accurately measured globally from satellites. Before that time, the data came from tide gauges at coastal stations around the world. Both the satellite measurements and tide-gauge observations show that the rate of sea level rise has accelerated. Statistical analysis reveals that the rate of rise is closely correlated with temperature; the warmer it gets, the faster the sea level rises (Rahmstorf, 2007).

Sea level rise is an inevitable consequence of global warming for two main reasons: ocean water expands as it heats up, and additional water flows into the oceans from the ice that melts on land. For the period 1961–2003, thermal expansion contributed ~40% to the observed sea level rise, while shrinking mountain glaciers and ice sheets have contributed ~60% (Domingues et al., 2008).

Sea level has risen faster than expected (Rahmstorf et al., 2007) (see Fig. 11.1). The average rate of rise for 1993–2008, as measured from

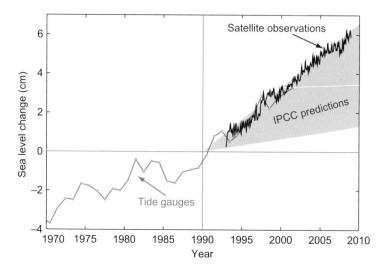

FIGURE 11.1   Sea level change during 1970–2010. The tide gauge data are indicated in red (Church & White, 2006) and satellite data in blue (Cazenave *et al.*, 2009). The grey band shows the projections of the IPCC TAR for comparison.

satellite, is 3.4 mm per year (Cazenave *et al.*, 2009), while the IPCC Third Assessment Report (TAR) projected a best estimate of 1.9 mm per year for the same period. Actual rise has thus been 80% faster than projected by models. (Note that the more recent models of the 2007 IPCC report still project essentially the same sea level rise as those of the TAR, to within 10%.)

Future sea level rise is highly uncertain, as the mismatch between observed and modeled sea level already suggests. The main reason for the uncertainty is in the response of the big ice sheets of Greenland and Antarctica.

Sea level is likely to rise much more by 2100 than the often-cited range of 18–59 cm from the IPCC AR4. As noted in the IPCC AR4, the coupled models used in developing the 21st century sea level projections did not include representations of dynamic ice sheets. As such, the oft-cited 18–59 cm projected sea level rise only included simple mass balance estimates of the sea level contribution from the Greenland and Antarctic ice sheets. As a consequence of an assumed positive mass balance over the Antarctic ice sheet in the AR4, Antarctica was estimated to have contributed to global sea level *decline* during the 21st century in that report. However, the Antarctic ice sheet is currently losing mass as a consequence of dynamical processes (see Fig. 7.3). Based on a number of new studies, the synthesis document of the 2009 Copenhagen Climate Congress (Richardson *et al.*, 2009) concluded that "updated estimates of

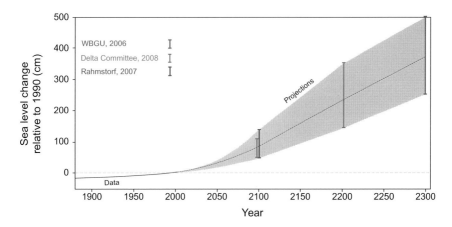

FIGURE 11.2   Some recent projections of future sea level rise. Future projections are from Rahmstorf (2007) and WBGU (2006), while those projections represented here as "Delta Committee" are from Vellinga *et al.* (2008).

the future global mean sea level rise are about double the IPCC projections from 2007."

Sea level will continue to rise for many centuries after global temperature is stabilized, since it takes that much time for the oceans and ice sheets to fully respond to a warmer climate. Some recent estimates of future rise are compiled in Fig. 11.2. These estimates highlight the fact that unchecked global warming is likely to raise sea level by several meters in coming centuries, leading to the loss of many major coastal cities and entire island states.

12

Abrupt Change and Tipping Points

KEY POINTS

- There are several tipping elements in the climate system that could pass a tipping point this century due to human activities, leading to abrupt and/or irreversible change.

- 1 °C global warming (above 1980–1999) carries moderately significant risks of passing large-scale tipping points, and 3 °C global warming would give substantial or severe risks.

- There are prospects for early warning of approaching tipping points, but if we wait until a transition begins to be observed, in some cases it would be unstoppable.

WHAT IS A TIPPING POINT?

A *tipping point* is a critical threshold at which the future state of a system can be qualitatively altered by a small change in forcing (Lenton *et al.*, 2008; Schellnhuber, 2009). A *tipping element* is a part of the Earth system (at least sub-continental in scale) that has a tipping point (Lenton *et al.*, 2008). Policy-relevant tipping elements are those that could be forced past a tipping point this century by human activities. *Abrupt climate change* is the subset of tipping point change, which occurs faster than its cause. Tipping point change also includes transitions that are slower than their cause (in both cases the rate is determined by the system itself). In either case, the change in state may be reversible or irreversible. *Reversible* means that when the forcing is returned below the tipping point, the system recovers its original state, either abruptly or gradually. *Irreversible* means that it does not (it takes a larger change in forcing to recover). Reversibility in principle does not mean that changes will be reversible in practice. A tipping element may lag anthropogenic forcing such that once a transition begins to be observed, a much larger change in state is already inevitable.

ARE THERE TIPPING POINTS IN THE EARTH'S CLIMATE SYSTEM?

There are a number of tipping points in the climate system, based on the understanding of its non-linear dynamics, and as revealed by past abrupt climate changes and model behavior (Pitman & Stouffer, 2006; Schellnhuber, 2009). Some models pass tipping points in future projections, and recent observations show abrupt changes already underway in the Arctic. Recent work has identified a shortlist of potential policy-relevant tipping elements in the climate system that could pass a tipping point this century and undergo a transition this millennium under projected climate change (Lenton *et al.*, 2008). These are shown with some other candidates in Fig. 12.1.

WHICH ONES ARE OF THE GREATEST CONCERN? HOW HAS THIS BEEN ASSESSED?

The tipping points of greatest concern are those that are the nearest (least avoidable) and those that have the largest negative impacts. Generally, the more rapid and less reversible a transition is, the greater its impacts. Additionally, any amplifying feedback to global climate change may increase concern, as can interactions, whereby tipping one element encourages tipping another. The proximity of some tipping points has been assessed through expert elicitation (Kriegler *et al.*, 2009; Lenton *et al.*, 2008). Proximity, rate, and reversibility have been also assessed through literature review (Lenton *et al.*, 2008), but there is a need for more detailed consideration of impacts. Some of the most concerning regions and their tipping elements are now discussed.

Arctic: The Greenland ice sheet (GIS) may be nearing a tipping point where it is committed to shrink (Kriegler *et al.*, 2009; Lenton *et al.*, 2008). Striking amplification of seasonal melt was observed in 2007, associated with record Arctic summer sea ice loss (Mote, 2007). Once underway, the transition to a smaller Greenland ice cap will have low reversibility, although it is likely to take several centuries (and is therefore not abrupt). The impacts via sea level rise will ultimately be large and global, but will depend on the rate of ice sheet shrinkage.

Antarctic: The West Antarctic ice sheet (WAIS) is currently assessed to be further from a tipping point than the GIS, but this is more uncertain

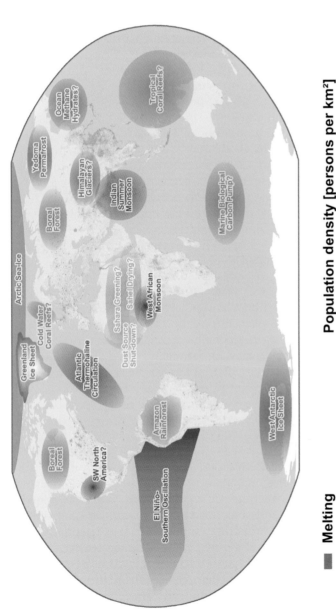

FIGURE 12.1 Map of potential policy-relevant tipping elements in the Earth climate system overlain on global population density. Tipping elements shown in blue involve melting of large masses of ice, those shown in red involve changes in atmospheric and oceanic circulation, and those shown in green involve loss of unique biomes. Question marks indicate systems whose status as policy-relevant tipping elements is particularly uncertain. Figure by V Huber, TM Lenton, and HJ Schellnhuber, adapted from Lenton et al. (2008).

(Kriegler *et al.*, 2009; Lenton *et al.*, 2008). The WAIS has the potential for more rapid change and hence greater impacts. The loss of ice shelves around the Antarctic Peninsula, such as Larsen B, followed by the acceleration of glaciers they were buttressing, highlights a mechanism that could threaten parts of the WAIS. The main East Antarctic ice sheet (EAIS) is thought to be more stable than the WAIS. However, there is evidence that changes are taking place along its marine sector, which drains more ice than all of West Antarctica.

Amazonia: The Amazon rainforest experienced widespread drought in 2005 turning the region from a sink to a source (0.6–0.8 Gt C per year) of carbon (Phillips *et al.*, 2009). If anthropogenic-forced lengthening of the dry season continues (Vecchi *et al.*, 2006), and droughts increase in frequency or severity (Cox *et al.*, 2008), the rainforest could reach a tipping point resulting in dieback of up to ~80% of the rainforest (Cook & Vizy, 2008; Cox *et al.*, 2004; Salazar *et al.*, 2007; Scholze *et al.*, 2006), and its replacement by savannah. This could take a few decades, would have low reversibility, large regional impacts, and knock-on effects far away. Widespread dieback is expected in a >4 °C warmer world (Kriegler *et al.*, 2009), and it could be committed to at a lower global temperature, long before it begins to be observed (Jones *et al.*, 2009).

West Africa: The Sahel and West African Monsoon (WAM) have experienced rapid but reversible changes in the past, including devastating drought from the late 1960s through the 1980s. Forecasts of future weakening of the Atlantic thermohaline circulation contributing to "Atlantic Niño" conditions, including strong warming in the Gulf of Guinea (Cook & Vizy, 2006), could disrupt the seasonal onset of the WAM (Chang *et al.*, 2008) and its later "jump" northwards (Hagos & Cook, 2007) into the Sahel. Perversely, if the WAM circulation collapses, this could lead to wetting of parts of the Sahel as moist air is drawn in from the Atlantic to the West (Cook & Vizy, 2006; Patricola & Cook, 2008), greening the region in what would be a rare example of a positive tipping point.

India: The Indian Summer Monsoon is probably already being disrupted (Meehl *et al.*, 2008; Ramanathan *et al.*, 2005) by an atmospheric brown cloud haze that sits over the sub-continent and, to a lesser degree, the Indian Ocean. This haze is comprised of a mixture of soot, which absorbs sunlight, and some reflecting sulfate. It causes heating of the atmosphere rather than the land surface, weakening the seasonal establishment of a land–ocean temperature gradient, which is critical in triggering monsoon onset (Ramanathan *et al.*, 2005). In some future projections, brown cloud haze forcing could lead to a doubling of drought frequency within a decade (Ramanathan *et al.*, 2005) with large impacts, although transitions should be highly reversible.

Several other candidate tipping elements and mechanisms could become a major concern, for example, carbon loss from permafrost.

Recently it has been suggested that a region of permafrost known as the Yedoma, which stores up to ~500 Gt C (Zimov *et al.*, 2006) could be tipped into irreversible breakdown driven by internal biochemical heat generation (Khvorostyanov *et al.*, 2008a, 2008b). However, the tipping point is estimated to be relatively distant.

HOW DO TIPPING POINTS RELATE TO AMPLIFYING FEEDBACKS ON CLIMATE CHANGE?

Tipping points are often confused with the phenomenon of amplifying feedbacks on climate change. All tipping elements must have some strong amplifying feedback — detailed elsewhere (Lenton *et al.*, 2008) — in their own internal or regional climate dynamics in order to exhibit a threshold, but they need not have an amplifying feedback to global climate change. Tipping elements that could have an amplifying feedback to global climate change include the Amazon rainforest (dieback would make it a CO_2 source, which could ultimately release up to ~100 Gt C), the thermohaline circulation (weakening or collapse would lead to net out-gassing of CO_2), and the Yedoma permafrost (release of up to ~500 Gt C). Tipping elements that could have a diminishing feedback on global climate change include boreal forest (dieback would release CO_2 but this would be outweighed by cooling due to increased land surface albedo from unmasked snow cover; Betts, 2000) and the Sahel/Sahara (greening would take up CO_2 and probably increase regional cloud cover).

SHOULD WE BE CONCERNED ABOUT GLOBAL AMPLIFYING FEEDBACKS?

Amplifying feedbacks from individual tipping elements are mostly fairly weak at the global scale. However, other (non-tipping element) amplifying feedbacks, including a potential future switch in the average response of the land biosphere from a CO_2 sink to a CO_2 source, could significantly amplify CO_2 rise and global temperature on the century timescale (Friedlingstein *et al.*, 2006). The Earth's climate system is already in a state of strong amplifying feedback from relatively fast physical climate responses (Bony *et al.*, 2006) (e.g., water vapor feedback). In any system with strong amplifying feedback, relatively small additional feedbacks can have a disproportionate impact on the global state (in this case, temperature), because of the non-linear way in which amplifiers work together.

IS THERE A GLOBAL TIPPING POINT?

A global tipping point can only occur if a net amplifying feedback becomes strong enough to produce a threshold, whereby the global system is committed to a change in state, carried by its own internal dynamics. Despite much talk in the popular media about such "runaway" climate change, there is as yet no strong evidence that the Earth as a whole is near such a threshold. Instead, "amplified" climate change is a much better description of what we currently observe and project for the future.

WHICH ANTHROPOGENIC FORCING AGENTS ARE DANGEROUS?

The total cumulative emissions of CO_2 (and other long-lived green-house gases) determine long-term committed climate changes and hence the fate of those tipping elements that are sensitive to global mean temperature change, are slow to respond, and/or have more distant thresholds. Key examples are the large ice sheets (GIS and WAIS). Uneven sulfate (Rotstayn & Lohmann, 2002) and soot (Ramanathan & Carmichael, 2008; Ramanathan et al., 2005) aerosol forcing are most dangerous for monsoons. Soot deposition on snow and ice (Flanner et al., 2007; Ramanathan & Carmichael, 2008) is a key danger to Arctic tipping elements as it is particularly effective at forcing melting (Flanner et al., 2007). Increasing soot aerosol, declining sulfate aerosol (Shindell & Faluvegi, 2009), and increasing short-lived greenhouse gases (Hansen et al., 2007) (methane and tropospheric ozone) have also contributed to rapid Arctic warming, and together far outweigh the CO_2 contribution. The current mitigation of SO_2 emissions and hence sulfate aerosol is a mixed blessing for climate tipping elements; it may, for example, be benefiting the Sahel region (Rotstayn & Lohmann, 2002) but endangering the Amazon (Cox et al., 2008) and the Arctic sea ice (Shindell & Faluvegi, 2009). Land cover change may also drive large areas of continents from being relatively robust to climate change to being highly vulnerable.

IS THERE ANY PROSPECT FOR EARLY WARNING OF AN APPROACHING TIPPING POINT?

Recent progress has been made in identifying and testing generic potential early warning indicators of an approaching tipping point (Dakos, 2008; Livina & Lenton, 2007; Lenton et al., 2008, 2009; Scheffer et al., 2009). Slowing down in response to perturbation is a nearly universal property of systems approaching various types of tipping point (Dakos *et al.*, 2008; Scheffer *et al.*, 2009). This has been successfully detected in past climate records approaching different transitions (Dakos *et al.*, 2008; Livina & Lenton 2007), and in model experiments (Dakos *et al.*, 2008; Livina & Lenton, 2007; Lenton *et al.*, 2009). Flickering between states may also occur prior to a more permanent transition (Bakke *et al.*, 2009). Other early warning indicators are being explored for ecological tipping points (Biggs *et al.*, 2009), including increasing variance (Biggs *et al.*, 2009), skewed responses (Biggs *et al.*, 2009; Guttal & Jayaprakash, 2008) and their spatial equivalents (Guttal & Jayaprakash, 2009). These could potentially be applied to anticipating climate tipping points.

13

Lessons from the Past

KEY POINTS

- The reconstruction of past climate reveals that the recent warming observed in the Arctic, and in the Northern Hemisphere in general, is anomalous in the context of natural climate variability over the last 2000 years.

- New ice-core records confirm the importance of greenhouse gases for past temperatures on earth, and show that CO_2 levels are higher now than they have ever been during the last 800,000 years.

RECONSTRUCTING THE LAST TWO MILLENNIA

Knowledge of climate during past centuries can help us to understand natural climate change and put modern climate change into context. There have been a number of studies to reconstruct trends in global and hemispheric surface temperature over the last millennium (Esper *et al.*, 2002; Mann *et al.*, 1998; Moberg *et al.*, 2005), all of which show recent Northern Hemisphere warmth to be anomalous in the context of at least the past millennium, and likely longer (Jansen *et al.*, 2007). The first of these reconstructions has come to be known as the "hockey stick" reconstruction (Mann *et al.*, 1998, 1999). Some aspects of the hockey stick reconstruction were subsequently questioned, e.g. whether the 20th century was the warmest at a hemispheric average scale (Soon & Baliunas, 2003), and whether the reconstruction is reproducible, or verifiable (McIntyre & McKitrick, 2003), or might be sensitive to the method used to extract information from tree ring records (McIntyre & McKitrick, 2005a, 2005b). Whilst these criticisms have been rejected in subsequent work (e.g., Jansen *et al.*, 2007; Rutherford *et al.*, 2005; Wahl *et al.*, 2006; Wahl & Ammann, 2007) the US National Research Council convened a committee to examine the state of the science of reconstructing the climate of the past

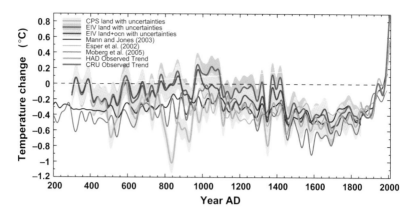

FIGURE 13.1   Comparison of various Northern Hemisphere temperature reconstructions, with estimated 95% confidence intervals shown (from Mann *et al.*, 2008. Copyright National Academy of Sciences, USA). Republished with permission of American Geophysical Union, from "Arctic sea ice decline: Faster than forecast", Stroeve et al, Geophysical Research Letters Vol34 L09501, 2007; permission conveyed through Copyright Clearance Center, Inc. From "Proxy-based reconstructions of hemispheric and global surface temperature variations over the past two millennia," Mann et al, PNAS vol. 105 no. 36 13252-13257, Copyright 2008 National Academy of Sciences, U.S.A.

millennium. The NRC report published in 2006 largely supported the original findings of Mann *et al.* (1998, 1999) and recommended a path toward continued progress in this area (NRC, 2006).

Mann *et al.* (2008) addressed the recommendations of the NRC report by reconstructing surface temperature at a hemispheric and global scale for much of the last 2000 years using a greatly expanded data set for decadal-to-centennial climate changes, along with recently updated instrumental data and complementary methods that have been thoroughly tested and validated with climate model simulations. Their results extend previous studies and conclude that recent Northern Hemisphere surface temperature increases are likely anomalous in a long-term context (Fig. 13.1).

Kaufman *et al.* (2009) independently concluded that recent Arctic warming is without precedent in at least 2000 years (Fig. 13.2), reversing a long-term millennial-scale cooling trend caused by astronomical forcing (i.e., orbital cycles). Warmth during the peak of the "Medieval Climate Anomaly" of roughly AD 900–1100 may have rivaled modern warmth for certain regions such as the western tropical Pacific (Oppo *et al.*, 2009), and some regions neighboring the North Atlantic (Mann *et al.*, 2009b). However, such regional warming appears to reflect a redistribution of warmth by changes in atmospheric circulation, and is generally offset by cooling elsewhere (e.g., the eastern and central tropical Pacific) to yield hemispheric and global temperatures that are lower than those of recent decades.

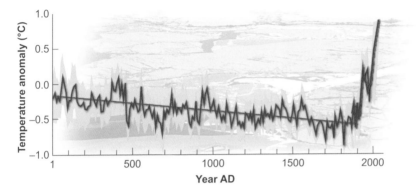

FIGURE 13.2   Blue line: estimates of Arctic air temperatures over the last 2000 years based on proxy records from lake sediments, ice cores, and tree rings. The green line shows the best fit long-term cooling trend for the period ending 1900. The red line shows the recent warming based on actual observations. (Courtesy *Science*, modified by the University Corporation for Atmospheric Research). From "Recent Warming Reverses Long-Term Arctic Cooling," Kaufman et al, *Science* 4 September 2009: 1236-1239. *Reprinted with permission from AAAS.*

Ice Core Records of Greenhouse Gases

Changes in past atmospheric CO_2 and CH_4 concentrations can be determined by measuring the composition of air trapped in ice cores and through the analyses of leaf stomata density and geochemical analyses of marine sediment cores.

The Dome Concordia (Dome C) ice core CO_2 and CH_4 records, drilled by the European Project for Ice Coring in Antarctica (EPICA), were published in 2004 and 2005 detailing events back to 440,000 years and 650,000 years respectively (EPICA community members, 2004; Siegenthaler *et al.*, 2005). In 2008, the record was extended to 800,000 years (Loulergue *et al.*, 2008; Lüthi *et al.*, 2008). The newly extended records reveal that current greenhouse gas levels (~390ppm) are at least 40% higher than at any time over the past 800,000 years. We must travel back at least two to three million years, and perhaps as far as fifteen million years, to the Pliocene and Miocene epochs of geological time to find equivalent greenhouse gas levels in the atmosphere (Haywood *et al.*, 2007; Kürschner *et al.*, 1996; Raymo *et al.*, 1996; Tripati *et al.*, 2009).

Strong correlations of CH_4 and CO_2 with temperature reconstructions are maintained throughout the new 800,000-year record (Loulergue *et al.*, 2008; Lüthi *et al.*, 2008). Temperature warming typically comes before increases in atmospheric CO_2 over the ice-core record. This finding is consistent with the view that natural CO_2 variations constitute a feedback in the glacial–interglacial cycle rather than a primary cause (Shackleton, 2000), something that has recently been explained in detail with the help of climate model experiments (Ganopolski & Roche, 2009). Changes in the

Earth's orbit around the Sun are the pacemaker for glacial–interglacial cycles (Berger, 1978; Hays *et al.*, 1976), but these rather subtle orbital changes must be amplified by climate feedbacks in order to explain the large differences in global temperature and ice volume, and the relative abruptness of the transitions between glacial and interglacial periods (Berger *et al.*, 1998; Clark *et al.*, 1999).

Palaeo Constraints on Climate and Earth System Sensitivity

One of the key questions for climate research is to determine how sensitively the Earth's climate responds to a given change in our planet's radiation budget. This is often described by "climate sensitivity," defined as the equilibrium global temperature response to a doubling of atmospheric CO_2 concentration.

IPCC AR4 summarizes the research aimed at characterizing the uncertainty in climate sensitivity (e.g., Andronova & Schlesinger, 2001; Annan & Hargreaves, 2006; Frame *et al.*, 2005) by stating that "climate sensitivity is likely to lie in the range 2 °C to 4.5 °C, with a most likely value of about 3 °C." More recent studies have agreed with this assessment (e.g., Knutti & Hegerl, 2008). These estimates of climate sensitivity have also been used to determine the likely impacts, both environmental and social/economic, of various CO_2 stabilization scenarios, or the level of greenhouse gas emissions consistent with stabilization of the global mean temperature below a certain value (e.g., Meinshausen *et al.*, 2009; see section "Mitigating global warming" in Chapter 14).

Q & A

Isn't Climate Always Changing, Even Without Human Interference?

Of course! But past climate changes are no cause for complacency; indeed, they tell us that the Earth's climate is very sensitive to changes in forcing. Two main conclusions can be drawn from climate history:

(1) Climate has always responded strongly if the radiation balance of the Earth is disturbed. That suggests the same will happen again, now that humans are altering the radiation balance by increasing greenhouse gas concentrations. In fact, data from climate changes in the Earth's history have been used to quantify how strongly a given change in the radiation balance alters the global temperature (i.e., to determine the *climate sensitivity*). The data confirm that our climate system is as sensitive as our climate models suggest, perhaps even more so.

(2) Impacts of past climate changes have been severe. The last great Ice Age, when it was globally 4–7 °C colder than now, completely transformed the Earth's surface and its ecosystems, and sea level was 120 m lower. When the Earth last was 2–3 °C warmer than now, during the Pliocene three million years ago, sea level was 25–35 m higher due to the smaller ice sheets present in the warmer climate.

Despite the large natural climate changes, the recent global warming does stick out already. Climate reconstructions suggest that over the past two millennia, global temperature has never changed by more than 0.5 °C in a century and has never been as warm as at present (e.g., Mann *et al.*, 2008; and references therein).

Q & A

Are We Just in a Natural Warming Phase, Recovering from the "Little Ice Age"?

No. A "recovery" of climate is not a scientific concept, since the climate does not respond like a pendulum that swings back after it was pushed in one direction. Rather, the climate responds like a pot of water on the stove: it can only get warmer if you add heat, according to the most fundamental law of physics, conservation of energy. The Earth's heat budget (its *radiation balance*) is well understood. By far the biggest change in the radiation balance over the past 50 years, during which three quarters of global warming has occurred, is due to the human-caused increase in greenhouse gas concentrations (see above). Natural factors have had a slightly cooling effect during this period.

Q & A *(cont'd)*

Global temperatures are now not only warmer than in the 16th–19th centuries, sometimes dubbed "the Little Ice Age" (although this term is somewhat misleading in that this largely regional phenomenon has little in common with real ice ages). Temperatures are in fact now globally warmer than any time in the past 2000 years—even warmer than in the "medieval optimum" a thousand years ago (see Fig. 13.1). This is a point that all global climate reconstructions by different groups of researchers, based on different data and methods, agree upon.

Q & A

In Climate History, didn't CO_2 Change in Response to Temperature, Rather than the Other Way Round?

It works both ways: CO_2 changes affect temperature due to the greenhouse effect, while temperature changes affect CO_2 concentrations due to the carbon cycle response. This is what scientists call a feedback loop.

(Continued)

Q & A   *(cont'd)*

If global temperatures are changed, the carbon cycle will respond (typically with a delay of centuries). This can be seen during the ice age cycles of the past three million years, which were caused by variations in the Earth's orbit (the so-called Milankovich cycles). The CO_2 feedback amplified and globalized these orbital climate changes: without the lowered CO_2 concentrations and reduced greenhouse effect, the full extent of ice ages cannot be explained, nor can the fact that the ice ages occurred simultaneously in both hemispheres. The details of the lag-relationship of temperature and CO_2 in Antarctic records have recently been reproduced in climate model experiments (Ganopolski and Roche, 2009) and they are entirely consistent with the major role of CO_2 in climate change. During the warming at the end of ice ages, CO_2 was released from the oceans—just the opposite of what we observe today, where CO_2 is increasing in both the ocean and the atmosphere.

If the CO_2 concentration in the atmosphere is changed, then the temperature follows because of the greenhouse effect. This is what is happening now that humans release CO_2 from fossil sources. But this has also happened many times in Earth's history. CO_2 concentrations have changed over millions of years due to natural carbon cycle changes associated with plate tectonics (continental drift), and climate has tracked those CO_2 changes (e.g., the gradual cooling into ice-age climates over the past 50 million years).

A rapid carbon release, not unlike what humans are causing today, has also occurred at least once in climate history, as sediment data from 55 million years ago show. This "Paleocene—Eocene thermal maximum" brought a major global warming of $\sim 5\ °C$, a detrimental ocean acidification, and a mass extinction event. It serves as a stark warning to us today.

CHAPTER

14

The Future

KEY POINTS

- Global mean air temperature is projected to warm 2 °C–7 °C above pre-industrial levels by 2100. The wide range is mainly due to uncertainty in future emissions.

- There is a very high risk of the warming exceeding 2 °C unless global emissions peak and start to decline rapidly by 2020.

- Warming rates will accelerate if positive carbon feedbacks significantly diminish the efficiency of the land and ocean to absorb our CO_2 emissions.

- Many indicators are currently tracking near or above the worst-case projections from the IPCC AR4 set of model simulations.

CLIMATE PROJECTIONS

There has been no new coordinated set of future climate model projections undertaken since the IPCC AR4. Instead, much of the new research over the past few years has focused on preparation for the next round of IPCC simulations for AR5, and continued evaluation of the AR4 model runs. This includes new analyses of the observed rate of climate change in comparison to the IPCC AR4 projections (e.g., Rahmstorf 2007; Stroeve et al. 2007), and new calculations that take existing simulations and incorporate coupled carbon feedbacks and other processes (e.g., Allen et al., 2009; Zickfeld et al., 2009). While models exhibit good skill at capturing the mean present-day climate, some recent observed changes, notably sea level rise and Arctic sea ice melt, are occurring at a faster rate than anticipated by IPCC AR4. This is a cause for concern as it suggests that some amplifying feedbacks and missing processes, such as land ice melt, are occurring faster than first predicted.

The latest estimates of global mean air temperature projected out to 2100 are shown in Fig. 14.1. The wide range in the projection envelope is

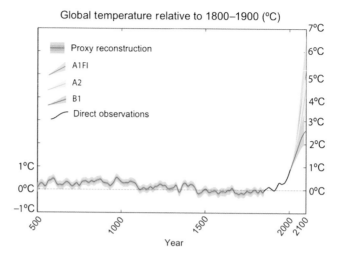

FIGURE 14.1 Reconstructed global-average temperature relative to 1800–1900 (blue) and projected global-average temperature out to 2100 (the latter from IPCC AR4 SPM). The envelopes B1, A2, and A1FI refer to the IPCC AR4 projections using those scenarios. The reconstruction is taken from Mann *et al.* (2008).

primarily due to uncertainty in future emissions. At the high end of emissions, with business as usual for several decades to come, global mean warming is estimated to reach 4–7 °C by 2100, locking in climate change at a scale that would profoundly and adversely affect all of human civilization and all of the world's major ecosystems. At the lower end of emissions, something that would require urgent, deep, and long-lasting cuts in fossil fuel use, and active preservation of the world's forests, global mean warming is projected to reach 2–3 °C by the century's end. While clearly a better outcome than the high emissions route, global mean warming of even just 1.5–2.0 °C still carries a significant risk of adverse impacts on ecosystems and human society. For example, 2 °C global temperature rise could lead to sufficient warming over Greenland to melt much of its ice sheet (Oppenheimer & Alley, 2005), raising sea level by over 6 m and displacing hundreds of millions of people worldwide.

Despite the certainty of a long-term warming trend in response to rising greenhouse gases, there is no expectation that the warming will be monotonic and follow the emissions pathway on a year-to-year basis. This is because natural variability and the 11-year solar cycle, as well as sporadic volcanic eruptions, generate short-term variations super-imposed on the long-term trend (Lean & Rind, 2009). Even under a robust century-long warming trend of around 4 °C, we still expect to see the temperature record punctuated by isolated but regular 10-year periods of no trend, or even modest cooling (Easterling & Wehner, 2009). Such decades therefore do not spell the end of global warming—emissions must peak and decline well before that is to occur. In fact, the peak in global temperature might not be reached until several centuries after emissions peak (e.g., Allen *et al.*, 2009). Even after emissions stop completely, and there has been a stabilization of global warming, atmospheric temperatures are not expected to decline much for many centuries to millennia (Eby *et al.*, 2009; Matthews & Caldeira, 2008; Solomon *et al.*, 2009) because of the long lifetime of CO_2 in the atmosphere. Furthermore, dry season rainfall reductions in several regions are expected to become irreversible (Solomon *et al.*, 2009).

MITIGATING GLOBAL WARMING

While global warming can be stopped, it cannot easily be reversed due to the long lifetime of carbon dioxide in the atmosphere (Eby *et al.*, 2009; Solomon *et al.*, 2009). Even a thousand years after reaching a zero-emission society, temperatures will remain elevated, likely cooling down by only a few tenths of a degree below their peak values. Therefore, decisions taken now have profound and practically irreversible consequences for many generations to come, unless affordable ways to extract CO_2 from the atmosphere in massive amounts can be found in the future. The chances of this do not appear to be promising.

The temperature at which global warming will finally stop depends primarily on the *total amount* of CO_2 released to the atmosphere since industrialization (Allen *et al.*, 2009; Meinshausen *et al.*, 2009; Zickfeld *et al.*, 2009). This is again due to the long lifetime of atmospheric CO_2. If global warming is to be stopped, global CO_2 emissions must therefore eventually decline to zero. The sooner emissions stop, the lower the final warming will be. From a scientific point of view, a cumulative CO_2 budget for the world would thus be a natural element of a climate policy agreement. Such an agreed global budget could then be distributed amongst countries, for example on the basis of equity principles (e.g., WBGU, 2009).

The most widely supported policy goal is to limit global warming to at most 2 °C above the pre-industrial temperature level (often taken for example as the average 19th century temperature, although the exact definition does not matter much due to the small variations in pre-industrial temperatures). Many nations have publically recognized the importance of this 2 °C limit. Furthermore, the group of least developed countries as well as the 43 small island states (AOSIS) are calling for limiting global warming to only 1.5 °C. The Synthesis Report of the Copenhagen climate congress (Richardson *et al.*, 2009), the largest climate science conference of 2009, concluded that "temperature rises above 2 °C will be difficult for contemporary societies to cope with, and are likely to cause major societal and environmental disruptions through the rest of the century and beyond."

A number of recent scientific studies have investigated in detail what global emissions trajectories would be compatible with limiting global warming to 2 °C. The answer has to be given in terms of probabilities, to reflect the remaining uncertainty in the climate response to elevated CO_2, and the uncertainty in the stability of carbon stored in the land and ocean systems. Meinshausen *et al.* (2009) found that if a total of 1000 Gt of CO_2 is emitted for the period 2000–2050, the likelihood of exceeding the 2-degree warming limit is around 25%. In 2000–2009, about 350 Gt have already been emitted, leaving only 650 Gt for 2010–2050. At current emission rates this budget would be used up within 20 years.

An important consequence of the rapidly growing emissions rate, and the need for a limited emissions budget, is that any delay in reaching the peak in emissions drastically increases the required rapidity and depth of future emission cuts (see Fig. 14.2 and also England *et al.*, 2009). In Fig. 14.2, emissions in the green exemplary path are 4 Gt CO_2 in the year

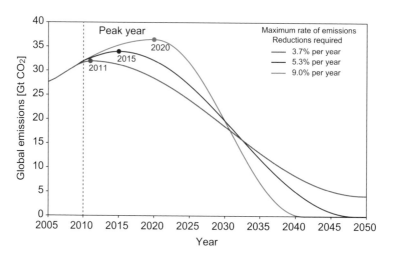

FIGURE 14.2   Examples of global emission pathways where cumulative CO_2 emissions equal 750 Gt during the time period 2010–2050 (1 Gt C = 3.67 Gt CO_2). At this level, there is a 67% probability of limiting global warming to a maximum of 2 °C. The graph shows that the later the peak in emissions is reached, the steeper their subsequent reduction has to be. The figure shows variants of a global emissions scenario with different peak years: 2011 (green), 2015 (blue), and 2020 (red). In order to achieve compliance with these curves, maximum annual reduction rates of 3.7% (green), 5.3% (blue), or 9.0% (red) would be required (relative to 2008). *Source: German Advisory Council on Global Change; WBGU, 2009.* From "Factsheet 2/2009 Climate change: Why 2°C?" Figure 2; German Advisory Council on Global Change WBGU, Berlin, 2009.

2050, which, with a projected world population of around nine billion, would leave only less than half-a-ton per person per year. While the exact number will depend strongly on the path taken, the required decline in emissions combined with a growing population will mean that by 2050, annual per capita CO_2 emissions very likely will need to be below 1 ton.

Although CO_2 is the most important anthropogenic climate forcing, other greenhouse gases as well as aerosols also play a non-negligible role. Successful limitation of the non-CO_2 climate forcing would therefore create more leeway in the allowable CO_2 emissions budget. Studies have shown that attractive options for particularly rapid and cost-effective climate mitigation are the reduction of black carbon (soot) pollution and tropospheric low-level ozone (Wallack & Ramanathan, 2009). In contrast to CO_2, these are very short-lived gases in the atmosphere, and therefore respond rapidly to policy measures.

Bibliography

Åkerman, H. J., & Johansson, M. (2008). Thawing permafrost and thicker active layers in sub-arctic Sweden. *Permafrost and Periglacial Processes, 19*, 279–292.

Alexander, L. V., & Arblaster, J. M. (2009). Assessing trends in observed and modelled climate extremes over Australia in relation to future projections. *International Journal of Climatology, 29*, 417–435.

Allan, R. P., & Soden, B. J. (2008). Atmospheric warming and the amplification of precipitation extremes. *Science, 321*, 1481–1484.

Allen, R. J., & Sherwood, S. C. (2008). Warming maximum in the tropical upper troposphere deduced from thermal winds. *Nature Geoscience, 1*, 399–403.

Allen, M. R., Frame, D. J., Huntingford, C., Jones, C. D., Lowe, J. A., Meinshausen, M., et al. (2009). Warming caused by cumulative carbon emissions toward the trillionth tonne. *Nature, 458*, 1163–1166.

Alley, R. B., Marotzke, J., Nordhaus, W. D., Overpeck, J. T., Peteet, D. M., Pielke, R. A., Jr., et al. (2003). Abrupt climate change. *Science, 299*, 2005–2010.

Allison, I., Alley, R. B., Fricker, H. A., Thomas, R. H., & Warner, R. (2009). Ice sheet mass balance and sea level. *Antarctic Science, 21*, 413–426.

Andronova, N., & Schlesinger, M. E. (2001). Objective estimation of the probability distribution for climate sensitivity. *Journal of Geophysical Research, 106*, 22605–22612.

Annan, J. D., & Hargreaves, J. C. (2006). Using multiple observationally-based constraints to estimate climate sensitivity. *Geophysical Research Letters, 33*, L06704.

Archer, D., Buffett, B., & Brovkin, V. (2009). Ocean methane hydrates as a slow tipping point in the global carbon cycle. *Proceedings of the National Academy of Sciences, 106*(49), 20596–20601.

Arzel, O., Fichefet, T., & Goosse, H. (2006). Sea ice evolution over the 20th and 21st centuries as simulated by the current AOGCMs. *Ocean Modelling, 12*, 401–415.

Aumann, H. H., Ruzmaikin, A., & Teixeira, J. (2008). Frequency of severe storms and global warming. *Geophysical Research Letters, 35*, L19805.

Bahr, D. B., Dyurgerov, M. B., & Meier, M. F. (2009). Sea-level rise from glaciers and ice caps: a lower bound. *Geophysical Research Letters, 36*, L03501.

Bakke, J., Lie, Ø., Heegaard, E., Dokken, T., Haug, G. H., Birks, H. H., et al. (2009). Rapid oceanic and atmospheric changes during the Younger Dryas cold period. *Nature Geoscience, 2*, 202–205.

Bala, G., Caldeira, K., & Wickett, M. (2007). Combined climate and carbon-cycle effects of large-scale deforestation. *Proceedings of the National Academy of Sciences, 104*, 6550–6555.

Bamber, J. L., Riva, R. E. M., Vermeersen, B. L. A., & LeBrocq, A. M. (2009). Reassessment of the potential sea-level rise from a collapse of the West Antarctic Ice Sheet. *Science, 324*(5929), 901–903.

Barnett, T. P., Pierce, D. W., Hidalgo, H. G., Bonfils, C., Santer, B. D., Das, T., et al. (2008). Human induced changes in the hydrology of the western United States. *Science, 319*, 1080–1083.

Barrett, B. E., Nicholls, K. W., Murray, T., Smith, A. M., & Vaughan, D. G. (2009). Rapid recent warming on Rutford Ice Stream, West Antarctica, from Borehole thermometry. *Geophysical Research Letters, 36*, L02708.

Benestad, R. E., & Schmidt, G. A. (2009). Solar trends and global warming. *Journal of Geophysical Research, 114*, D14101.

Berger, A. (1978). Long-term variations of daily insolation and quaternary climatic changes. *Journal of the Atmospheric Sciences, 35*, 2362–2367.

Berger, A., & Loutre, M. F. (1991). Insolation values for the climate of the last 19 million years. *Quaternary Science Reviews, 10*, 297–317.

Berger, A., Loutre, M. F., & Gallée, H. (1998). Sensitivity of the LLN climate model to the astronomical and CO_2 forcings over the last 200 ky. *Climate Dynamics, 14*, 615–629.

Betts, R. A. (2000). Offset of the potential carbon sink from boreal afforestation by decreases in surface Albedo. *Nature, 408*, 187–190.

Betts, R. A., Boucher, O., Collins, M., Cox, P. M., Falloon, P. D., Gedney, N., et al. (2007). Projected increases in continental river runoff due to plant responses to carbon dioxide. *Nature, 448*, 1037–1041.

Biggs, R., Carpenter, S. R., & Brock, W. A. (2009). Turning back from the brink: detecting an impending regime shift in time to avert it. *Proceedings of the National Academy of Sciences, 106*(3), 826–831.

Bindoff, N. L., Willebrand, J., Artale, V., Cazenave, A., Gregory, J., Gulev, S., et al. (2007). Observations: oceanic climate change and sea level. In S. Solomon, et al. (Eds.), *Climate change 2007: The Physical Science Basis. Contribution of working group I to the fourth assessment report of the intergovernmental panel on climate change.* Cambridge, United Kingdom and New York, NY, USA: Cambridge University Press.

Bondeau, A., Smith, P. C., Zaehle, S., Schaphoff, S., Lucht, W., Cramer, W., et al. (2007). Modelling the role of agriculture for the 20th century global terrestrial carbon balance. *Global Change Biology, 13*, 679–706.

Bony, S., Dufresne, J.-L., Colman, R., Kattsov, V. M., Allan, R. P., Bretherton, C. S., et al. (2006). How well do we understand and evaluate climate change feedback processes? *Journal of Climate, 19*, 3445–3482.

Booth, B. B., Jones, C. D., Collins, M., Totterdell, I., Cox, P., Sitch, S., et al. (2009). *Global warming uncertainties due to carbon cycle feedbacks exceed those due to CO_2 emissions.* Paper presented at European Geosciences Union General Assembly. Vienna: Copernicus Publications.

Braun, M., & Humbert, A. (2009). Recent retreat of Wilkins ice shelf reveals new insights in ice shelf breakup mechanisms. *IEEE Geoscience and Remote Sensing Letters, 6*(2), 263–267.

Brewer, P. G. (2009). A changing ocean seen with clarity. *Proceedings of the National Academy of Sciences, 106*(30), 12213–12214.

Brook, E., Archer, D., Dlugokencky, E., Frolking, S., & Lawrence, D. (2008). Potential for abrupt changes in atmospheric methane, in abrupt climate change. A report by the U.S. Climate Change Science Program and the Subcommittee on Global Change Research. In J. P. McGeehin (Ed.) (pp. 360–452). U.S. Geological Survey.

Canadell, J. G., Le Quéré, C., Raupach, M. R., Field, C. B., Buitenhuis, E. T., Ciais, P., et al. (2007). Contributions to accelerating atmospheric CO_2 growth from economic activity, carbon intensity, and efficiency of natural sinks. *Proceedings of the National Academy of Sciences, 104*(47), 18866–18870.

Cavalieri, D. J., & Parkinson, C. L. (2008). Antarctic sea ice variability and trends, 1979-2006. *Journal Of Geophysical Research, 113*, C07004.

Cazenave, A., Dominh, K., Guinehut, S., Berthier, E., Llovel, W., Ramillien, G., et al. (2009). Sea level budget over 2003-2008: a reevaluation from GRACE space gravimetry, satellite altimetry and ARGO. *Global and Planetary Change, 65*, 83–88.

CCSP. (2008a). *Weather and climate extremes in a changing climate. Regions of focus: North America, Hawaii, Caribbean, and U.S. Pacific Islands.* A Report by the U.S. Climate Change Science Program. Washington, D.C., USA: Department of Commerce, NOAA's National Climatic Data Center. 164 pp.

CCSP. (2008b). *Abrupt climate change. A report by the U.S. climate change science program and the subcommittee on global change research.* Reston, VA: U.S. Geological Survey. 459 pp.

Chang, P., Zhang, R., Hazeleger, W., Wen, C., Wan, X., Ji, L., et al. (2008). Oceanic link between abrupt change in the North Atlantic Ocean and the African monsoon. *Nature Geoscience, 1,* 444–448.

Chapman, W. L., & Walsh, J. E. (2007). A synthesis of Antarctic temperatures. *Journal of Climate, 20,* 4096–4117.

Chen, J., Wilson, C., Blankenship, D., & Tapley, B. (2006). Antarctic mass rates from GRACE33. *Geophysical Research Letters, 33,* L11502.

Church, J. A., & White, N. J. (2006). A 20th century acceleration in global sea-level rise. *Geophysical Research Letters, 33,* L01602.

Chylek, P., & Lohmann, U. (2008). Aerosol radiative forcing and climate sensitivity deduced from the Last Glacial Maximum to Holocene transition. *Geophysical Research Letters, 35,* L04804.

Clark, P. U., Alley, R. A., & Pollard, D. (1999). Northern Hemisphere ice-sheet influences on global climate change. *Science, 286,* 1104–1111.

Cogley, J. G. (2009). Geodetic and direct mass-balance measurements: comparison and joint analysis. *Annals of Glaciology, 50,* 96–100.

Comiso, J. C., & Nishio, F. (2008). Trends in the sea ice cover using enhanced and compatible AMSR-E, SSM/I, and SMMR data. *Journal of Geophysical Research, 113,* C02S07.

Cook, A., Fox, A. J., Vaughan, D. G., & Ferrigno, J. G. (2005). Retreating glacier-fronts on the Antarctic Peninsula over the last 50 years. *Science, 22,* 541–544.

Cook, K. H., & Vizy, E. K. (2006). Coupled model simulations of the West African Monsoon system: twentieth- and twenty-first-century simulations. *Journal of Climate, 19,* 3681–3703.

Cook, K. H., & Vizy, E. K. (2008). Effects of twenty-first-century climate change on the Amazon Rain Forest. *Journal of Climate, 21,* 542–560.

Cox, P. M., Betts, R. A., Collins, M., Harris, P. P., Huntingford, C., & Jones, C. D. (2004). Amazonian forest dieback under climate-carbon cycle projections for the 21st century. *Theoretical and Applied Climatology, 78,* 137–156.

Cox, P. M., & Jones, C. D. (2008). Data from the past illuminates the modern dance of climate and carbon dioxide. *Science, 321,* 1642–1643.

Cox, P. M., Harris, P. P., Huntingford, C., Betts, R. A., Collins, M., Jones, C. D., et al. (2008). Increasing risk of Amazonian drought due to decreasing aerosol pollution. *Nature, 453,* 212–216.

Cruz, F., Pitman, A. J., & McGregor, J. (2010). Probabilistic simulations of the impact of increasing leaf-level atmospheric carbon dioxide on the global land surface. *Climate Dynamics, 34,* 361–379.

Cui, X., & Graf, H. F. (2009). Recent land cover changes on the Tibetan Plateau: a review. *Climatic Change, 94,* 47–61.

Curry, R., Dickson, B., & Yashayaev, I. (2003). A change in the freshwater balance of the Atlantic Ocean over the past four decades,. *Nature, 426,* 826–829.

Dakos, V., Scheffer, M., van Nes, E. H., Brovkin, V., Petoukhov, V., & Held, H. (2008). Slowing down as an early warning signal for abrupt climate change. *Proceedings of the National Academy of Sciences, 105*(38), 14308–14312.

Delworth, T. L., Clark, P. U., Holland, M., Johns, W. E., Kuhlbrodt, T., Lynch-Stieglitz, J., et al. (2008). The potential for abrupt change in the Atlantic meridional overturning circulation, in abrupt climate change. A report by the U.S. climate change science program and the subcommittee on global change research. In J. P. McGeehin (Ed.) (pp. 258–359). Reston, VA: U.S. Geological Survey.

Dessler, A. E., Zhang, Z., & Yang, P. (2008). Water-vapor climate feedback inferred from climate fluctuations, 2003–2008. *Geophysical Research Letters, 35,* L20704.

Domingues, C. M., Church, J. A., White, N. J., Gleckler, P. J., Wijffels, S. E., Barker, P. M., et al. (2008). Improved estimates of upper-ocean warming and multi-decadal sea-level rise. *Nature, 453,* 1090–1093.

Dorrepaal, E., S. Toet, Logtestijn, R. S. P.v., Swart, E., Weg, M. J.v. d., Callaghan, T. V., et al. (2009). Carbon respiration from subsurface peat accelerated by climate warming in the subarctic. *Nature, 460,* 616–619.

Easterling, D. R., & Wehner, M. F. (2009). Is the climate warming or cooling? *Geophysical Research Letters, 36,* L08706.

Eby, M., Zickfeld, K., Montenegro, A., Archer, D., Meissner, K. J., & Weaver, A. J. (2009). Lifetime of anthropogenic climate change: millennial time scales of potential CO_2 and surface temperature perturbations. *Journal of Climate, 22,* 2501–2511.

Eisenman, I., & Wettlaufer, J. S. (2009). Nonlinear threshold behavior during the loss of Arctic sea ice. *Proceedings of the National Academy of Sciences, 106*(1), 28–32.

Elsner, J. B., Kossin, J. P., & Jagger, T. H. (2008). The increasing intensity of the strongest tropical cyclones. *Nature, 455,* 92–95.

Emanuel, K., Sundararajan, R., & Williams, J. (2008). Hurricanes and global warming: results from downscaling IPCC AR4 simulations. *Bulletin of the American Meteorological Society, 89,* 347–367.

England, M. H., Gupta, A. S., & Pitman, A. J. (2009). Constraining future greenhouse gas emissions by a cumulative target. *Proceedings of the National Academy of Sciences, 106,* 16539–16540.

EPICA community members. (2004). Eight glacial cycles from an Antarctic ice core. *Nature, 429,* 623–628.

Esper, J., Cook, E. R., & Schweingruber, F. H. (2002). Low-frequency signals in long tree-ring chronologies for reconstructing past temperature variability. *Science, 295,* 2250–2253.

Fabry, V. J., Seibel, B. A., Feely, R. A., & Orr, J. C. (2008). Impacts of ocean acidification on marine fauna and ecosystem processes. *ICES Journal of Marine Science, 65,* 414–432.

Fargione, J., Hill, J., Tilman, E., Polasky, S., & Hawthorn, P. (2008). Land clearing and the biofuel debt. *Science, 319,* 1235–1238.

Fischer, E. M., Seneveritane, S. I., Lüthi, D., & Schär, C. (2007). Contribution of land-atmosphere coupling to recent European heat waves. *Geophysical Research Letters, 34,* L06707.

Flanner, M. G., Zender, C. S., Randerson, J. T., & Rasch, P. J. (2007). Present-day climate forcing and response from black carbon in snow. *Journal of Geophysical Research, 112,* D11202.

Frame, D. J., Booth, B. B. B., Kettleborough, J. A., Stainforth, D. A., Gregory, J. M., Collins, M., et al. (2005). Constraining climate forecasts: the role of prior assumptions. *Geophysical Research Letters, 32,* L09702.

Frederick, E., Manizade, S., Martin, C., Sonntag, J., Swift, R., Thomas, R. H., et al. (2004). Greenland ice sheet: increased coastal thinning. *Geophysical Research Letters, 31,* L24402.

Friedlingstein, P., Cox, P. M., Betts, R. A., Jones, C., von Bloh, W., Brovkin, V., et al. (2006). Climate-carbon cycle feedback analysis: results from the C4MIP model intercomparison. *Journal of Climate, 19,* 3337–3353.

Friedlingstein, P., Houghton, R. A., Marland, G., Hackler, J., Boden, T. A., Conway, T. J., et al. (2010). Update on CO_2 emissions. *Nature Geoscience, 3,* 811–812.

Fyke, J. G., & Weaver, A. J. (2006). The effect of potential future climate change on the marine methane hydrate stability zone. *Journal of Climate, 19,* 5903–5916.

Galloway, J. N., Townsend, A. R., Erisman, J. W., Bekunda, M., Cai, Z., Freney, J. R., et al. (2008). Transformation of the nitrogen cycle: recent trends, questions and potential solutions. *Science, 320,* 889–892.

Ganopolski, A., & Roche, D. M. (2009). On the nature of lead–lag relationships during glacial–interglacial climate transitions. *Quaternary Science Reviews, 28*(27-28), 3361–3378.

Gedney, N., Cox, P. M., Betts, R. A., Boucher, O., Huntingford, C., & Stott, P. A. (2006). Detection of a direct carbon dioxide effect in continental river runoff records. *Nature, 439*, 835–838.

Gleason, K. L., Lawrimore, J. H., Levinson, D. H., Karl, T. R., & Karoly, D. J. (2008). A revised U.S. climate extremes index. *Journal of Climate, 21*, 2124–2137.

Goosse, H., et al. (2009). Consistent past half-century trends in the atmosphere, the sea ice and the ocean at high southern latitudes. *Climate Dynamics, 33*, 999–1016.

Guan, D., Peters, G. P., Weber, C. L., & Hubacek, K. (2009). Journey to world top emitter: an analysis of the driving forces of China's recent CO_2 emissions surge. *Geophysical Research Letters, 36*, L04709.

Guttal, V., & Jayaprakash, C. (2008). Changing skewness: an early warning signal of regime shifts in ecosystems. *Ecology Letters, 11*, 450–460.

Guttal, V., & Jayaprakash, C. (2009). Spatial variance and spatial skewness: leading indicators of regime shifts in spatial ecological systems. *Theoretical Ecology, 2*, 3–12.

Hagos, S. M., & Cook, K. H. (2007). Dynamics of the West African Monsoon Jump. *Journal of Climate, 20*, 5264–5284.

Hall, D. K., Williams, R. S., Luthcke, S. B., & Digirolamo, N. E. (2008). Greenland ice sheet surface temperature, melt and mass loss: 2000-06. *Journal of Glaciology, 54*(184), 81–93.

Hanna, E., Huybrechts, P., Steffen, K., Cappelen, J., Huff, R., Shuman, C., et al. (2008). Increased runoff from melt from the Greenland ice sheet: a response to global warming. *Journal of Climate, 21*, 331–341.

Hanna, E., Cappelen, J., Fettweis, X., Huybrechts, P., Luckman, A., & Ribergaard, M. H. (2009). Hydrologic response of the Greenland ice sheet: the role of oceanographic forcing. *Hydrological Processes, 23*(1), 7–30.

Hansen, J., Sato, M., Ruedy, R., Kharecha, P., Lacis, A., Miller, R., et al. (2007). Dangerous human-made interference with climate: a GISS modelE study. *Atmospheric Chemistry and Physics, 7*(9), 2287–2312.

Hansen, B., & Østerhus, S. (2007). Faroe Bank channel overflow 1995-2005. *Progress in Oceanography, 75*, 817–856.

Harris, C., Arenson, L. U., Christiansen, H. H., Etzelmüller, B., Frauenfelder, R., Gruber, S., et al. (2009). Permafrost and climate in Europe: monitoring and modelling thermal, geomorphological and geotechnical responses. *Earth-Science Review, 92*(3-4), 117–171.

Hays, J. D., Imbrie, J., & Shackleton, N. J. (1976). Variations in the Earth's orbit: pacemaker of the ice ages. *Science, 194*, 1121–1132.

Haywood, A. M., Valdes, P. J., Hill, D. J., & Williams, M. (2007). The mid-Pliocene warm period: a test-bed for integrating data and models, in deep-time perspectives on climate change: marrying the signal from computer models and biological proxies. Special Publication, The Geological Society. In M. Williams, et al. (Eds.) (pp. 443–458). London: The Micropalaeontological Society.

Hock, R., Woul, M. D., Radic, V., & Dyurgerov, M. B. (2009). Mountain glaciers and ice caps around Antarctica make a large sea-level rise contribution. *Geophysical Research Letters, 36*, L07501.

Hofmann, M., & Schellnhuber, H. J. (2009). Oceanic acidification affects marine carbon pump and triggers extended marine oxygen holes. *Proceedings of the National Academy of Sciences, 106*, 3017–3022.

Hofmann, M., & Rahmstorf, S. (2009). On the stability of the Atlantic meridional overturning circulation. *Proceedings of the National Academy of Sciences, 106*(49), 20584–20589.

Holland, M. M., Bitz, C. M., & Tremblay, B. (2006). Future abrupt reductions in the summer Arctic sea ice. *Geophysical Research Letters, 33*, L23503.

Holland, D. M., Thomas, R. H., deYoung, B., Ribergaard, M. H., & Lyberth, B. (2008). Acceleration of Jakobshavn Isbrae triggered by warm subsurface ocean waters. *Nature Geoscience, 28*, 659–664.

House, J. I., Huntingford, C., Knorr, W., Cornell, S. E., Cox, P. M., Harris, G. R., et al. (2008). What do recent advances in quantifying climate and carbon cycle uncertainties mean for climate policy? *Environmental Research Letters, 3,* 044002.

Howat, I. M., Joughin, I., & Scambos, T. A. (2007). Rapid changes in ice discharge from Greenland outlet glaciers. *Science, 315,* 1559–1561.

Howat, I. M., Smith, B. E., Joughin, I., & Scambos, T. A. (2008). Rates of southeast Greenland ice volume loss from combined ICESat and ASTER observations. *Geophysical Research Letters, 35,* L17505.

Hoyos, C. D., Agudelo, P. A., Webster, P. J., & Curry, J. A. (2006). Deconvolution of the factors contributing to the increase in global hurricane intensity. *Science, 312,* 94–97.

Hyvönen, R., Ågren, G. I., G.I., Linder, S., Persson, T., Cotrufo, M. F., et al. (2007). The likely impact of elevated CO_2, nitrogen deposition, increased temperature and management on carbon sequestration in temperate and boreal forest ecosystems: a literature review. *New Phytologist, 173,* 463–483.

IPCC. (2001). *Climate change 2001: The scientific basis. contribution of working group I to the third assessment report of the intergovernmental panel on climate change (IPCC TAR).* Cambridge, UK and New York, NY, USA: Cambridge University Press. 881 pp.

IPCC. (2007). *Climate change 2007: The physical science basis. contribution of working group I to the fourth assessment report of the intergovernmental panel on climate change (AR4).* Cambridge, UK & New York, NY, USA: Cambridge University Press. 996 pp.

Jansen, E., Overpeck, J., Briffa, K. R., Duplessy, J.-C., Joos, F., Masson-Delmotte, V., et al. (2007). Palaeoclimate. In S. Solomon, et al. (Eds.), *Climate change 2007: The physical science basis. Contribution of working group I to the fourth assessment report of the intergovernmental panel on climate change.* Cambridge, United Kingdom and New York, NY, USA: Cambridge University Press.

Jin, H.-j., Yu, Q.-h., Wang, S.-l., & Lü, L.-z. (2008). Changes in permafrost environments along the Qinghai-Tibet engineering corridor induced by anthropogenic activities and climate warming. *Cold Regions Science and Technology, 53*(3), 317–333.

Johannessen, O., Khvorostovsky, K., Miles, M., & Bobylev, L. (2005). Recent ice-sheet growth in the interior of Greenland. *Science, 310,* 1013–1016.

Johnson, G. C., & Gruber, N. (2007). Decadal water mass variations along 20 W in the Northeastern Atlantic Ocean. *Progress in Oceanography, 73,* 277–295.

Johnson, G. C., Purkey, S. G., & Toole, J. M. (2008a). Reduced Antarctic meridional over-turning circulation reaches the North Atlantic Ocean. *Geophysical Research Letters, 35,* L22601.

Johnson, G. C., Purkey, S. G., & Bullister, J. L. (2008b). Warming and freshening in the abyssal southeastern Indian Ocean. *Journal of Climate, 21,* 5351–5363.

Jones, K. F., & Light, B. (2008). Sunlight, water, and ice: extreme Arctic sea ice melt during the summer of 2007. *Geophysical Research Letters, 35,* L11501.

Jones, G. S., Stott, P. A., & Christidis, N. (2008). Human contribution to rapidly increasing frequency of very warm Northern Hemisphere summers. *Journal of Geophysical Research, 113,* D02109.

Jones, C., Lowe, J., Liddicoat, S., & Betts, R. A. (2009). Commited ecosystem change due to climate change. *Nature Geoscience, 2,* 484–487.

Joos, F., & Spahni, R. (2008). Rates of change in natural and anthropogenic radiative forcing over the past 20,000 years. *Proceedings of the National Academy of Sciences, 105,* 1425–1430.

Kaser, G., Cogley, J. G., Dyurgerov, M. B., Meier, M. F., & Ohmura, A. (2006). Mass balance of glaciers and ice caps: consensus estimates for 1961-2004. *Geophysical Research Letters, 33,* L19501.

Kaufman, D. S., Schneider, D. P., McKay, N. P., Ammann, C. M., Bradley, R. S., Briffa, K. R., et al. (2009). Recent warming reverses long-term Arctic cooling. *Science, 325,* 1236–1239.

Kharin, V. V., Zwiers, F. W., Zhang, X., & Hegerl, G. C. (2007). Changes in temperature and precipitation extremes in the IPCC ensemble of global coupled model simulations. *Journal of Climate, 20,* 1419—1444.

Khvorostyanov, D. V., Ciais, P., Krinner, G., & Zimov, S. A. (2008a). Vulnerability of east Siberia's frozen carbon stores to future warming. *Geophysical Research Letters, 35,* L10703.

Khvorostyanov, D. V., Krinner, G., Ciais, P., Heimann, M., & Zimov, S. A. (2008b). Vulnerability of permafrost carbon to global warming. Part I: model description and the role of heat generated by organic matter decomposition. *Tellus B, 60B,* 250—264.

Knutson, T. R., Sirutis, J. J., Garner, S. T., Vecchi, G. A., & Held, I. M. (2008). Simulated reduction in Atlantic hurricane frequency under twenty-first-century warming conditions. *Nature Geoscience, 1,* 359—364.

Knutti, R., & Hegerl, G. C. (2008). The equilibrium sensitivity of the Earth's temperature to radiation changes. *Nature Geoscience, 1,* 735—743.

Krabill, W., Abdalati, W., Frederick, E., Manizade, S., Martin, C., Sonntag, J., et al. (2000). Greenland ice sheet: high-elevation balance and peripheral thinning. *Science, 289,* 428—430.

Krabill, W., Hanna, E., Huybrechts, P., Abdalati, W., Cappelen, J., Csatho, B., et al. (2004). Greenland ice sheet: increased coastal thinning. *Geophysical Research Letters, 31,* L24402.

Kriegler, E., Hall, J. W., Held, H., Dawson, R., & Schellnhuber, H. J. (2009). Imprecise probability assessment of tipping points in the climate system. *Proceedings of the National Academy of Sciences, 106*(13), 5041—5046.

Kürschner, W. M., Van der Burgh, J., Visscher, H., & Dilcher, D. L. (1996). Oak leaves as biosensors of late Neogene and early Pleistocene paleoatmospheric CO_2 concentrations. *Marine Micropaleontology, 27,* 299—312.

Kwok, R., & Rothrock, D. A. (2009). Decline in Arctic sea ice thickness from submarine and ICESat records: 1958—2008. *Geophysical Research Letters, 36,* L15501.

Lam, P., Lavik, G., Jensen, M. M., Vossenberg, J.v. d., Schmid, M., Woebken, D., et al. (2009). Revising the nitrogen cycle in the Peruvian oxygen minimum zone. *Proceedings of the National Academy of Sciences, 106,* 4752—4757.

Latif, M., & Keenlyside, N. S. (2009). El Niño/Southern Oscillation response to global warming. *Proceedings of the National Academy of Sciences, 106*(49), 20578—20583.

Lawrence, D. M., & Slater, A. G. (2005). A projection of severe near-surface permafrost degradation during the 21st century. *Geophysical Research Letters, 32,* L24401.

Lawrence, D. M., Slater, A. G., Tomas, R. A., Holland, M. M., & Deser, C. (2008). Accelerated Arctic land warming and permafrost degradation during rapid sea ice loss. *Geophysical Research Letters, 35,* L11506.

Le Quéré, C., Rödenbeck, C., Buitenhuis, E. T., Conway, T. J., Langenfelds, R., Gomez, A., et al. (2007). Saturation of the Southern Ocean CO_2 sink due to recent climate change. *Science, 316,* 1735—1738.

Le Quéré, C., Raupach, M. R., Canadell, J. G., Marland, G., & e. al. (2009). Trends in the sources and sinks of carbon dioxide. *Nature Geosciences, 2,* 831—836.

Lean, J. L., & Rind, D. H. (2008). How natural and anthropogenic influences alter global and regional surface temperatures: 1889 to 2006. *Geophysical Research Letters, 35,* L18701.

Lean, J. L., & Rind, D. H. (2009). How will Earth's surface temperature change in future decades? *Geophysical Research Letters, 36,* L15708.

Lefebvre, W., Goosse, H., Timmermann, R., & Fichefet, T. (2004). Influence of the southern annular mode on the sea-ice-ocean system. *Journal of Geophysical Research, 109,* C090005.

Lemke, P., Alley, R. B., Allison, I., Carrasco, J., Flato, G., Fujii, Y., et al. (2007). Observations: changes in snow, ice and frozen ground. In S. Solomon, et al. (Eds.), *Climate change 2007: The physical science basis. Contribution of working group I to the fourth assessment report of the intergovernmental panel on climate change.* Cambridge, United Kingdom and New York, NY, USA: Cambridge University Press.

Lenton, T. M., Held, H., Kriegler, E., Hall, J. W., Lucht, W., Rahmstorf, S., et al. (2008). Tipping elements in the earth's climate system. *Proceedings of the National Academy of Sciences, 105*(6), 1786−1793.

Lenton, T. M., Myerscough, R. J., Marsh, R., Livina, V. N., Price, A. R., & Cox, S. J. (2009). Using GENIE to study a tipping point in the climate system. *Philosophical Transactions of the Royal Society A, 367*(1890), 871−884.

Letenmaier, D. P., & Milly, P. C. D. (2009). Land water and sea level. *Nature Geoscience, 2,* 452−454.

Lindsay, R. W., Zhang, J., Schweiger, A., Steele, M., & Stern, H. (2009). Arctic sea ice retreat in 2007 follows thinning trend. *Journal of Climate, 22*(1), 165−175.

Livina, V., & Lenton, T. M. (2007). A modified method for detecting incipient bifurcations in a dynamical system. *Geophysical Research Letters, 34,* L03712.

Lockwood, M., & Fröhlich, C. (2007). Recent oppositely directed trends in solar climate forcings and the global mean surface air temperature. *Proceedings of the Royal Society, A, 463*(2086), 2447−2460.

Lockwood, M., & Fröhlich, C. (2008). Recent oppositely directed trends in solar climate forcings and the global mean surface air temperature. II. Different reconstructions of the total solar irradiance variation and dependence on response time scale. *Proceedings of the Royal Society, A, 464*(2094), 1367−1385.

Lombard, A., Cazenave, A., Le Traon, P. Y., Guinehut, S., & Cabanes, C. (2006). Perspectives on present-day sea level change. *Ocean Dynamics, 56*(5-6), 445−451.

Loulergue, L., Schilt, A., Spahni, R., Masson-Delmotte, V., Blunier, T., Lemieux, B., et al. (2008). Orbital and millennial-scale features of atmospheric CH_4 over the past 800,000 years. *Nature, 453,* 383−386.

Lovenduski, N., Gruber, N., & Doney, S. C. (2008). Toward a mechanistic understanding of the decadal trends in the Southern Ocean carbon sink. *Global Biogeochemical Cycles, 22,* GB3016.

Lunt, D. J., Haywood, A. M., Schmidt, G. A., Salzmann, U., & Dowsett, H. J. (2010). Earth System sensitivity inferred from Pliocene modelling and data. *Nature Geoscience, 3,* 60−64.

Luthcke, S. B., Zwally, H. J., Abdalati, W., Rowlands, D. D., Ray, R. D., Nerem, R. S., et al. (2006). Recent Greenland ice mass loss by drainage system from satellite gravity observations. *Science, 314,* 1286−1289.

Lüthi, D., Floch, M. L., Bereiter, B., Blunier, T., Barnola, J.-M., Siegenthaler, U., et al. (2008). High-resolution carbon dioxide concentration record 650,000-800,000 years before present. *Nature, 453,* 379−382.

Malhi, Y., Roberts, J. T., Betts, R. A., Killeen, T. J., Li, W., & Nobre, C. A. (2008). Climate change, deforestation and the fate of the Amazon. *Science, 319,* 169−172.

Malhi, Y., Aragão, L. E. O. C., Galbraith, D., Huntingford, C., Fisher, R., Zelazowski, P., et al. (2009). Exploring the likelihood and mechanism of a climate-change induced dieback of the Amazon rainforest. *Proceedings of the National Academy of Sciences, 106,* 20610−20615.

Mann, M. E., Bradley, R. S., & Hughes, M. K. (1998). Global-scale temperature patterns and climate forcing over the past six centuries. *Nature, 392,* 779−787.

Mann, M. E., Bradley, R. S., & Hughes, M. K. (1999). Northern hemisphere temperatures during the past millennium: inferences, uncertainties, and limitations. *Geophysical Research Letters, 26,* 759−762.

Mann, M. E., & Emanuel, K. A. (2006). Atlantic hurricane trends linked to climate change. *Trans. AGU. EOS, 87*(24), 233.

Mann, M. E., Sabbatelli, T. A., & Neu, U. (2007). Evidence for a modest undercount bias in early historical Atlantic tropical cyclone counts. *Geophysical Research Letters, 34,* L22707.

Mann, M. E., Zhang, Z., Hughes, M. K., Bradley, R. S., Miller, S. K., & Rutherford, S. (2008). Proxy-based reconstructions of hemispheric and global surface temperature variations over the past two millennia. *Proceedings of the National Academy of Sciences, 105,* 13252–13257.

Mann, M. E., Woodruff, J. D., Donnelly, J. P., & Zhang, Z. (2009a). Atlantic hurricanes and climate over the past 1,500 years. *Nature, 460,* 880–883.

Mann, M. E., Z.Z., Rutherford, S., Bradley, R. S., Hughes, M. K., Shindell, D., et al. (2009b). Global signatures and dynamical origins of the "little ice age" and "medieval climate anomaly". *Science, 326,* 1256–1260.

Marsh, P. T., Brooks, H. E., & Karoly, D. J. (2009). Preliminary investigation into the severe thunderstorm environment of Europe simulated by the community climate systems model 3. *Atmospheric Research, 93,* 607–618.

Matthews, H. D., & Caldeira, K. (2008). Stabilizing climate requires near zero emissions. *Geophysical Research Letters, 35,* L04705.

McIntyre, S., & McKitrick, R. (2003). Corrections to the Mann et al. (1998) proxy database and northern hemispheric average temperature series. *Energy and Environment, 14*(6), 751–771.

McIntyre, S., & McKitrick, R. (2005a). Hockey sticks, principal components, and spurious significance. *Geophysical Research Letters, 32,* L03710.

McIntyre, S., & McKitrick, R. (2005b). The M&M critique of the MBH98 Northern Hemisphere climate index: update and implications. *Energy and Environment, 16,* 69–99.

McNeil, B. I., & Matear, R. J. (2007). Climate change feedbacks on oceanic pH. *Tellus-B, 59B,* 191–198.

McNeil, B. I., & Matear, R. J. (2008). Southern Ocean acidification: a tipping point at 450-ppm atmospheric CO_2. *Proceedings of the National Academy of Sciences, 105*(48), 18860–18864.

Meehl, G. A., Washington, W. M., Ammann, C. A., Arblaster, J. M., Wigleym, T. M. L., & Tebaldi, C. (2004). Combinations of natural and anthropogenic forcings in twentieth-century climate. *Journal of Climate, 19,* 3721–3727.

Meehl, G. A., Stocker, T. F., Collins, W. D., Friedlingstein, P., Gaye, A. T., Gregory, J. M., et al. (2007a). Global climate projections. In S. Solomon, et al. (Eds.), *Climate change 2007: The physical science basis. contribution of working group I to the fourth assessment report of the intergovernmental panel on climate change.* Cambridge, United Kingdom and New York, NY, USA: Cambridge University Press.

Meehl, G. A., Arblaster, J. M., & Tebaldi, C. (2007b). Contributions of natural and anthro-pogenic forcing to changes in temperature extremes over the U.S. *Geophysical Research Letters, 34,* L19709.

Meehl, G. A., Arblaster, J. M., & Collins, W. D. (2008). Effects of black carbon aerosols on the Indian monsoon. *Journal of Climate, 21,* 2869–2882.

Meier, M. F., & Dyurgerov, M. B. (2007). Glaciers dominate eustatic sea-level rise in the 21st century. *Science, 317,* 1064–1067.

Meinshausen, M., Meinshausen, N., Hare, W., Raper, S. C. B., Frieler, K., Knutti, R., et al. (2009). Greenhouse-gas emission targets for limiting global warming to 2°C. *Nature, 458,* 1158–1162.

Mercado, L. M., Bellouin, N., Sitch, S., Boucher, O., Huntingford, C., Wild, M., et al. (2009). Impact of change in diffuse radiation on the global land carbon sink. *Nature, 458,* 1014–1017.

Metzl, N. (2009). Decadal increase of oceanic carbon dioxide in Southern Indian surface ocean waters (1991–2007). *Deep Sea Research Part II: Topical Studies in Oceanography, 56* (8-10), 607–619.

Moberg, A., Sonechkin, D. M., Holmgren, K., Datsenko, N. M., & Karlén, W. (2005). Highly variable Northern Hemisphere temperatures reconstructed from low- and high-resolution proxy data. *Nature, 433,* 613–617.

Monaghan, A. J., Bromwich, D. H., Chapman, W. L., & Comiso, J. C. (2008). Recent variability and trends of Antarctic near-surface temperature. *Journal of Geophysical Research, 113,* D04105.

Mote, T. L. (2007). Greenland surface melt trends 1973-2007: evidence of a large increase in 2007. *Geophysical Research Letters, 34,* L22507.

Moy, A. D., Howard, W. R., Bray, S. G., & Trull, T. W. (2009). Reduced calcification in modern Southern Ocean planktonic foraminifera. *Nature Geoscience, 2,* 276–280.

Nakicenovic, N., Alcamo, J., Davis, G., de Vries, B., Fenhann, J., Gaffin, S., et al. (2000). *IPCC special report on emissions scenarios.* Cambridge, UK: Cambridge University Press.

NASA Goddard Institute for Space Studies. (2009). http://data.giss.nasa.gov/gistemp/2008/.

Nghiem, S. V., Rigor, I. G., Perovich, D. K., Clemente-Colón, P., Weatherly, J. W., & Neumann, G. (2007). Rapid reduction of Arctic perennial sea ice. *Geophysical Research Letters, 34,* L19504.

Nicholls, R. J., Wong, P. P., Burkett, V. R., Codignotto, J. O., Hay, J. E., McLean, R. F., et al. (2007). Coastal systems and low-lying areas. In M. L. Parry, et al. (Eds.), *Climate change 2007: Impacts, adaptation and vulnerability. Contribution of working group II to the fourth assessment report of the intergovernmental panel on climate change* (pp. 315–356). Cambridge, UK and New York, NY, USA: Cambridge University Press.

NOAA. (2009). http://www.ncdc.noaa.gov/sotc/.

NRC National Research Council. (2006). *Surface temperature reconstructions for the Last 2,000 Years.* Washington, DC: National Academies Press.

NSDIC National Snow and Ice Data Center. (2009). http://nsidc.org/news/press/20091005_minimumpr.html.

Oerlemans, J., Dyurgerov, M., & van de Wal, R. S. W. (2007). Reconstructing the glacier contribution to sea-level rise back to 1850. *The Cryosphere, 1*(1), 59–65.

Oppenheimer, M., & Alley, R. B. (2005). Ice sheets, global warming, and article 2 of the UNFCCC. *Climatic Change, 68,* 257–267.

Oppo, D. W., Rosenthal, Y., & Linsley, B. K. (2009). 2,000-year-long temperature and hydrology reconstructions from the Indo-Pacific warm pool. *Nature, 460,* 1113–1116.

Orr, J. C., Fabry, V. J., Aumont, O., & Bopp, L. (2005). Anthropogenic ocean acidification over the twenty-first century and its impact on calcifying organisms. *Nature, 437,* 681–686.

Orr, J. C., Jutterström, S., Bopp, L., Anderson, L. G., Cadule, P., Fabry, V. J., et al. (2009). Amplified acidification of the Arctic Ocean. *IOP Conf. Series: Earth and Environmental Science, 6,* 462009.

Oschlies, A., Schulz, K. G., Riebesell, U., & Schmittner, A. (2008). Simulated 21st century's increase in oceanic suboxia by CO_2-enhanced biotic carbon export. *Global Biogeochemical Cycles, 22,* GB4008.

Pall, P., Allen, M. R., & Stone, D. A. (2007). Testing the Clausius-Clapeyron constraint on changes in extreme precipitation under CO_2 warming. *Climate Dynamics, 28,* 351–363.

Patricola, C. M., & Cook, K. H. (2008). Atmosphere/vegetation feedbacks: a mechanism for abrupt climate change over northern Africa. *Journal of Geophysical Research (Atmospheres), 113,* D18102.

Pearson, P. L., & Palmer, M. R. (2000). Middle Eocene seawater pH and atmospheric carbon dioxide concentrations. *Science, 284,* 1824–1826.

Pedersen, C. A., Roeckner, E., Lüthje, M., & Winther, J.-G. (2009). A new sea ice albedo scheme including melt ponds for ECHAM5 general circulation model. *Journal of Geophysical Research, 114,* D08101.

Perovich, D. K., Light, B., Eicken, H., Jones, K. F., Runciman, K., & Nghiem, S. V. (2007). Increasing solar heating of the Arctic Ocean and adjacent seas, 1979-2005: attibution and role in the ice-albedo feedback. *Geophysical Research Letters, 34,* L19505.

Petrenko, V. V., Smith, A. M., Brook, E. J., Lowe, D., Riedel, K., Brailsford, G., et al. (2009). $^{14}CH_4$ measurements in Greenland ice: investigating last glacial termination CH_4 sources. *Science, 324,* 506—508.

Pfeffer, W. T., Harper, J. T., & O'Neel, S. (2008). Kinematic constraints on glacier contributions to 21st-century sea-level rise. *Science, 321,* 1340—1343.

Phillips, O. L., Aragao, L. E. O. C., Lewis, S. L., Fisher, J. B., Lloyd, J., López-González, G., et al. (2009). Drought sensitivity of the Amazon rainforest. *Science, 323,* 1344—1347.

Piao, S., Friedlingstein, P., Ciais, P., de Noblet-Ducoudré, N., Labat, D., & Zaehle, S. (2007). Changes in climate and land-use have a larger direct impact than rising CO_2 on global river runoff records. *Proceedings of the National Academy of Sciences, 104,* 15242—15247.

Piekle, R. A., Sr., Adegoke, J., Beltrán-Przekurat, A., Hiemstra, C. A., Lin, J., Nair, U. S., et al. (2007). An overview of regional land-use and land-cover impacts on rainfall. *Tellus B, 59,* 587—601.

Pitman, A. J., & Stouffer, R. J. (2006). Abrupt change in climate and climate models. *Hydrology and Earth System Sciences, 10,* 903—912.

Pitman, A. J., Narisma, G. T., & McAneney, J. (2007). The impact of climate change on the risk of forest and grassland fires in Australia. *Climatic Change, 84,* 383—401.

Pitman, A. J., de Noblet-Ducoudré, N., Cruz, F. T., Davin, E. L., Bonan, G. B., Brovkin, V., et al. (2009). Uncertainties in climate responses to past land cover change: first results from the LUCID intercomparison study. *Geophysical Research Letters, 36,* L14814.

Pollard, D., & DeConto, R. M. (2009). Modelling West Antarctic ice sheet growth and collapse through the past five million years. *Nature, 458,* 329—332.

Polyakov, I. V., Bhatt, U. S., Colony, R. L., Simmons, H. L., Walsh, J. E., Alekseev, G. V., et al. (2004). Variability of the intermediate Atlantic water of the Arctic Ocean over the last 100 years. *Journal of Climate, 17,* 4485—4497.

Portmann, R. W., Solomon, S., & Hegerl, G. C. (2009). Linkages between climate change, extreme temperature and precipitation across the United States. *Proceedings of the National Academy of Sciences, 106,* 7324—7329.

Pritchard, H. D., & Vaughan, D. G. (2007). Widespread acceleration of tidewater glaciers on the Antarctic Peninsula. *Journal of Geophysical Research, 112,* F03S29.

Pritchard, H. D., Arthern, R. J., Vaughan, D. G., & Edwards, L. A. (2009). Extensive dynamic thinning on the margins of the Greenland and Antarctic ice sheets. *Nature, 461,* 971—975.

Rahmstorf, S., Crucifix, M., Ganopolski, A., Goosse, H., Kamenkovich, I., Knutti, R., et al. (2005). Thermohaline circulation hysteresis: a model intercomparison. *Geophysical Research Letters, 32,* L23605.

Rahmstorf, S., Cazenave, A., Church, J. A., Hansen, J. E., Keeling, R. F., Parker, D. E., et al. (2007). Recent climate observations compared to projections. *Science, 316,* 709.

Rahmstorf, S. (2007). A semi-empirical approach to projecting future sea-level rise. *Science, 315,* 368—370.

Ramanathan, V., Chung, C., Kim, D., Bettge, T., Buja, L., Kiehl, J. T., et al. (2005). Atmospheric brown clouds: impacts on South Asian climate and hydrological cycle. *Proceedings of the National Academy of Sciences, 102,* 5326—5333.

Ramanathan, V., & Carmichael, G. (2008). Global and regional climate changes due to black carbon. *Nature Geoscience, 1,* 221—227.

Raupach, M. R., Marland, G., Ciais, P., Le Quéré, C., Canadell, J. G., Klepper, G., et al. (2007). Global and regional drivers of accelerating CO_2 emissions. *Proceedings of the National Academy of Sciences, 104,* 10288—10293.

Raymo, M. E., Grant, B., Horowitz, M., & Rau, G. H. (1996). Mid-Pliocene warmth: stronger greenhouse and stronger conveyor. *Marine Micropaleontology, 27,* 313—326.

Rayner, N. A., Brohan, P., Parker, D. E., Folland, C. K., Kennedy, J. J., Vanicek, M., et al. (2006). Improved analyses of changes and uncertainties in sea surface temperature measured in situ since the mid-nineteenth century: the HadSST2 data set. *Journal of Climate, 19*, 446–469.

Reichstein, M., Ciais, P., Papale, D., Valentini, R., Running, S. W., Viovy, N., et al. (2007). Reduction of ecosystem productivity and respiration during the European summer 2003 climate anomaly: a joint flux tower, remote sensing and modelling analysis. *Global Change Biology, 13*, 634–651.

Repo, M. E., Susiluoto, S., Lind, S. E., Jokinen, S., Elsakov, V., Biasi, C., et al. (2009). Large N_2O emissions from cryoturbated peat soil in tundra. *Nature Geoscience, 2*, 189–192.

Richardson, K., Steffen, W., Schellnhuber, H. J., Alcamo, J., Barker, T., Kammen, D. M., et al. (2009). *Climate change: Global risks, challenges & decisions*. Synthesis Report of the Copenhagen Climate Congress. University of Copenhagen.

Riebesell, U., Körtzinger, A., & Oschlies, A. (2009). Sensitivities of marine carbon fluxes to ocean change. *Proceedings of the National Academy of Sciences, 106*(49), 20602–20609.

Rigby, M., Prinn, R. G., Fraser, P. J., Simmonds, P. G., Langenfelds, R. L., Huang, J., et al. (2008). Renewed growth of atmospheric methane. *Geophysical Research Letters, 35*, L22805.

Rignot, E., Casassa, G., Gogineni, P., Krabill, W., Rivera, A., & Thomas, R. (2004). Accelerated ice discharge from the Antarctic Peninsula following the collapse of Larsen B ice shelf. *Geophysical Research Letters, 31*, L18401.

Rignot, E. (2006). Changes in ice dynamics and mass balance of the Antarctic ice sheet. *Philosophical Transactions of the Royal Society A, 364*(1844), 1637–1655.

Rignot, E., & Kanagaratnam, P. (2006). Changes in the velocity structure of the Greenland Ice Sheet. *Science, 311*, 986–990.

Rignot, E. (2008). Changes in West Antarctic ice stream dynamics observed with ALOS PALSAR data. *Geophysical Research Letters, 35*, L12505.

Rignot, E., Box, J. E., Burgess, E., & Hanna, E. (2008a). Mass balance of the Greenland ice sheet from 1958 to 2007. *Geophysical Research Letters, 35*, L20502.

Rignot, E., Bamber, J., van den Broeke, M., Davis, C., Li, Y., van de Berg, W., et al. (2008b). Recent Antarctic ice mass loss from radar interferometry and regional climate modelling. *Nature Geoscience, 1*, 106–110.

Rintoul, S. R. (2007). Rapid freshening of Antarctic Bottom Water formed in the Indian and Pacific oceans. *Geophysical Research Letters, 34*, L06606.

Rohling, E. J., Hemlebenrant, K. G. C., Siddall, M., Hoogakker, B. A. A., Bolshaw, M., & Kucera, M. (2008). High rates of sea-level rise during the last interglacial period. *Nature Geoscience, 1*, 38–42.

Rosa, R., & Seibel, B. A. (2008). Synergistic effects of climate-related variables suggest future physiological impairment in a top oceanic predator. *Proceedings of the National Academy of Sciences, 105*, 20776–20780.

Rotstatyn, L. D., & Lohmann, U. (2002). Tropical rainfall trends and the indirect aerosol effect. *Journal of Climate, 15*, 2103–2116.

Royer, D. L., Berner, R. A., & Park, J. (2007). Climate sensitivity constrained by CO_2 concentrations over the past 420 million years. *Nature, 446*, 530–532.

Rutherford, S., Mann, M. E., Osborn, T. J., Briffa, K. R., Jones, P. D., Bradley, R. S., et al. (2005). Proxy-based northern hemisphere surface temperature reconstructions: sensitivity to method, predictor network, target season, and target domain. *Journal of Climate, 18*(13), 2308–2329.

Sabine, C. L., Feely, R. A., Gruber, N., Key, R. M., Lee, K., Bullister, J. L., et al. (2004). The oceanic sink for anthropogenic CO_2. *Science, 305*, 367–371.

Salazar, L. F., Nobre, C. A., & Oyama, M. D. (2007). Climate change consequences on the biome distribution in tropical South America. *Geophysical Research Letters, 34*, L09708.

Santer, B. D., Mears, C., Wentz, F. J., Taylor, K. E., Gleckler, P. J., Wigley, T. M. L., et al. (2007). Identification of human-induced changes in atmospheric moisture content. *Proceedings of the National Academy of Sciences, 104*, 15248–15253.

Saunders, M. A., & Lea, A. S. (2008). Large contribution of sea surface warming to recent increase in Atlantic hurricane activity. *Nature, 451,* 557—560.

Scambos, T. A., Bohlander, J. A., Shuman, C. A., & Skvarca, P. (2004). Glacier acceleration and thinning after ice shelf collapse in the Larsen B embayment, Antarctica. *Geophysical Research Letters, 31,* L18402.

Scheffer, M., Bascompte, J., Brock, W. A., Brovkin, V., Carpenter, S. R., Dakos, V., et al. (2009). Early-warning signals for critical transitions. *Nature, 461,* 53—59.

Schellnhuber, H. J. (2009). Tipping elements in the Earth System. *Proceedings of the National Academy of Sciences, 106,* 20561—20563.

Scholze, M., Knorr, W., Arnell, N. W., & Prentice, I. C. (2006). A climate-change risk analysis for world ecosystems. *Proceedings of the National Academy of Sciences, 103*(35), 13116—13120.

Schuster, U., Watson, A. J., Bates, N., Corbière, A., Gonzaalez-Davila, M., Metzl, N., et al. (2009). Trends in North Atlantic sea surface pCO_2 from 1990 to 2006. *Deep Sea Research Part II: Topical Studies in Oceanography, 56,* 620—629.

Schuur, E. A. G., Bockheim, J., Canadell, J., Euskirchen, E., Field, C. B., Goryachkin, S. V., et al. (2008). Vulnerability of permafrost carbon to climate change: implications for the global carbon cycle. *BioScience, 58,* 701—714.

Shackleton, N. J. (2000). The 100,000-year ice-age cycle identified and found to lag temperature, carbon dioxide, and orbital eccentricity. *Science, 289,* 1897—1902.

Sheffield, J., & Wood, E. F. (2008). Global trends and variability in soil moisture and drought characteristics, 1950—2000, from observation-driven simulations of the terrestrial hydrologic cycle. *Journal of Climate, 21,* 432—458.

Shepherd, A., & Wingham, D. (2007). Recent sea-level contributions of the Antarctic and Greenland ice sheets. *Science, 315*(5818), 1529—1532.

Shindell, D., & Faluvegi, G. (2009). Climate response to regional radiative forcing during the twentieth century. *Nature Geoscience, 2,* 294—300.

Siegenthaler, U., Stocker, T. F., Monnin, E., Lüthi, D., Schwander, J., Stauffer, B., et al. (2005). Stable carbon cycle-climate relationship during the late Pleistocene. *Science, 310,* 1313—1317.

Sitch, S., Cox, P. M., Collins, W. J., & Huntingford, C. (2007). Indirect raditive forcing due to ozone effects on the land carbon sink. *Nature, 448,* 791—794.

Smith, J. B., Schneider, S. H., Oppenheimer, M., Yohe, G. W., Hare, W., Mastrandrea, M. D., et al. (2009). Assessing dangerous climate change through an update of the Intergovernmental Panel on Climate Change (IPCC) "reasons for concern". *Proceedings of the National Academy of Sciences, 106*(11), 4133—4137.

Sokolov, A. P., Stone, P. H., Forest, C. E., Prinn, R., Sarofim, M. C., Webster, M., et al. (2009). Probabilistic forecast for 21st century climate based on uncertainties in emissions (without policy) and climate parameters. *Journal of Climate, 22*(19), 5175—5204.

Solomon, S., Plattner, G. K., Knutti, R., & Friedlingstein, P. (2009). Irreversible climate change due to carbon dioxide emissions. *Proceedings of the National Academy of Sciences, 106,* 1704—1709.

Soon, W., & Baliunas, S. (2003). Proxy climatic and environmental changes of the past 1000 years. *Climate Research, 23,* 89—110.

Stammerjohn, S. E., Martinson, D. G., Smith, R. C., Yuan, X., & Rind, D. (2008). Trends in Antarctic annual sea ice retreat and advance and their relation to El Niño-Southern Oscillation and Southern Annular Mode variability. *Journal of Geophysical Research, 113,* C03S90.

Steffen, K., Clark, P. U., Cogley, J. G., Holland, D. M., Marshall, S., Rignot, E., et al. (2008). Rapid changes in glaciers and ice sheets and their impacts on sea level. In J. P. McGeehin (Ed.), *Abrupt climate change: a report by the U.S. climate change science program and the subcommittee on global change research* (pp. 60–142). Reston, VA: U.S. Geological Survey.

Steig, E. J., Schneider, D. P., Rutherford, S. D., Mann, M. E., Comiso, J. C., & Shindell, D. T. (2009). Warming of the Antarctic ice-sheet surface since the 1957 International Geophysical Year. *Nature, 457,* 459–462.

Stott, P. A., Sutton, R. T., & Smith, D. M. (2008). Detection and attribution of Atlantic salinity changes. *Geophysical Research Letters, 35,* L21702.

Stramma, L., Johnson, G. C., Sprintall, J., & Mohrholz, V. (2008). Expanding oxygen-minimum zones in the tropical oceans. *Science, 320,* 655–658.

Stroeve, J., Holland, M. M., Meier, W., Scambos, T. A., & Serreze, M. (2007). Arctic sea ice decline: faster than forecast. *Geophysical Research Letters, 34,* L09501.

Takahashi, T., Sutherland, S. C., Wanninkhof, R., Sweeney, C., Feely, R. A., Chipman, D. W., et al. (2009). Climatological mean and decadal changes in surface ocean pCO_2, and net sea-air CO_2 flux over the global oceans. *Deep Sea Research Part II: Topical Studies in Oceanography, 56*(8-10), 554–577.

Tarnocai, C., Canadell, J. G., Schuur, E. A. G., Kuhry, P., Mazhitova, G., & Zimov, S. (2009). Soil organic carbon pools in the northern circumpolar permafrost region. *Global Biogeochemical Cycles, 23,* GB2023.

Thomas, H., Friederike Prowe, A. E., Lima, I. D., Doney, S. C., Wanninkhof, R., Greatbatch, R. J., et al. (2008). Changes in the North Atlantic oscillation influence CO_2 uptake in the North Atlantic over the past 2 decades. *Global Biogeochemical Cycles, 22,* GB4027.

Thompson, D. W. J., & Solomon, S. (2002). Interpretation of recent Southern Hemisphere climate change. *Science, 296,* 895–899.

Thorne, P. W. (2008). The answer is blowing in the wind. *Nature Geoscience, 1,* 347–348.

Trapp, R. J., Diffenbaugh, N. S., Brooks, H. E., Baldwin, M. E., Robinson, E. D., & Pal, J. S. (2007). Changes in severe thunderstorm environment frequency during the 21st century caused by anthropogenically enhanced global radiative forcing. *Proceedings of the National Academy of Sciences, 104,* 19719–19723.

Trapp, R. J., Diffenbaugh, N. S., & Gluhovsky, A. (2009). Transient response of severe thunderstorm forcing to elevated greenhouse gas concentrations. *Geophysical Research Letters, 36,* L01703.

Tripati, A. K., Roberts, C. D., & Eagle, R. A. (2009). Coupling of CO2 and Ice sheet stability over major climate transitions of the last 20 million years. *Science, 326*(5958), 1394–1397.

Turner, J., Comiso, J. C., Marshall, G. J., Lachlan-Cope, T. A., Bracegirdle, T., Maksym, T., et al. (2009). Non-annular atmospheric circulation change induced by stratospheric ozone depletion and its role in the recent increase of Antarctic sea ice extent. *Geophysical Research Letters, 36,* L08502.

van den Broeke, M. (2005). Strong surface melting preceded collapse of Antarctic Peninsula ice shelf. *Geophysical Research Letters, 32,* L12815.

Vecchi, G. A., Soden, B. J., Wittenberg, A. T., Held, I. M., Leetmaa, A., & Harrison, M. J. (2006). Weakening of tropical Pacific atmospheric circulation due to anthropogenic forcing. *Nature, 441,* 73–76.

Vecchi, G. A., Swanson, K. L., & Soden, B. J. (2008). Whither hurricane activity? *Science, 322,* 687–689.

Velicogna, I., & Wahr, J. (2006). Acceleration of Greenland ice mass loss in spring 2004. *Nature, 443,* 329–331.

Velicogna, I. (2009). Increasing rates of ice mass loss from the Greenland and Antarctic ice sheets revealed by GRACE. *Geophysical Research Letters, 36,* L19503.

Vellinga, P., Katsman, C. A., Sterl, A., Beersma, J. J., Hazeleger, W., Church, J., et al. (2008). *Exploring high-end climate change scenarios for flood protection of the*

Netherlands: An international scientific assessment. Wageningen, The Netherlands: KNMI/Alterra.

Veron, J. E. N., Hoegh-Guldberg, O., Lenton, T. M., Lough, J. M., Obura, D. O., Pearce-Kelly, P., et al. (2009). The coral reef crisis: the critical importance of <350ppm CO_2. *Marine Pollution Bulletin, 58*, 1428−1437.

Wahl, E. R., Ritson, D. M., & Ammann, C. M. (2006). Comment on "Reconstructing past climate from noisy data". *Science, 312*, 529.

Wahl, E. R., & Ammann, C. M. (2007). Robustness of the Mann, Bradley, Hughes reconstruction of Northern Hemisphere surface temperatures: examination of criticisms based on the nature and processing of proxy climate evidence. *Climatic Change, 85*(1-2), 33−69.

Wallack, J. S., & Ramanathan, V. (2009). The other climate changers: why black carbon and ozone also matter. *Foreign Affairs, 88*(5), 105−113.

WBGU - German Advisory Council on Global Change. (2006). *The Future oceans - warming up, rising high, turning sour*. Berlin: WBGU. 110 pp.

WBGU - German Advisory Council on Global Change. (2009). *Solving the climate dilemma: The budget approach*. Berlin: WBGU. 58 pp.

Weart, S., & Pierrehumbert, R. T. (2007). http://www.realclimate.org/index.php/archives/2007/06/a-saturated-gassy-argument/.

Wentz, F. J., Ricciardulli, L., Hilburn, K., & Mears, C. (2007). How much more rain will global warming bring? *Science, 317*, 233−235.

Westerling, A. L., Hidalgo, H. G., Cayan, D. R., & Swetn, T. W. (2006). Warming and earlier spring increase Western U.S. forest wildfire activity. *Science, 313*, 940−943.

Wild, M., Ohmura, A., & Makowski, K. (2007). The impact of global dimming and brightening on global warming. *Geophysical Research Letters, 34*, L04702.

Wingham, D., Shepherd, A., Muir, A., & Marshall, G. (2006). Mass balance of the Antarctic ice sheet. *Transactions of the Royal Society of London Series A, 364*, 1627−1635.

Wouters, B., Chambers, D., & Schrama, E. J. O. (2008). GRACE observes small-scale mass loss in Greenland. *Geophysical Research Letters, 35*, L20501.

Yeh, S.-W., Kug, J.-S., Dewitte, B., Kwon, M.-H., Kirtman, B. P., & Jin, F.-F. (2009). El Niño in a changing climate. *Nature, 461*, 511−514.

Yin, J., Schlesinger, M. E., & Stouffer, R. J. (2009). Model projections of rapid sea-level rise on the northeast coast of the United States. *Nature Geoscience, 2*, 262−266.

Zhang, X., Zwiers, F. W., Hegerl, G. C., Lambert, F. H., Gillett, N. P., Solomon, S., et al. (2007). Detection of human influence on twentieth-century precipitation trends. *Nature, 448*, 461−465.

Zickfeld, K., Eby, M., Matthews, H. D., & Weaver, A. (2009). Setting cumulative emissions targets to reduce the risk of dangerous climate change. *Proceedings of the National Academy of Sciences, 106*, 16129−16134.

Zimov, S. A., Schuur, E. A. G., & Chapin, F. S. (2006). Permafrost and the global carbon budget. *Science, 312*, 1612−1613.

About the Authors

Ian Allison
Ian Allison is leader of the Ice Ocean Atmosphere and Climate program in the Australian Antarctic Division, a lead author of the IPCC Fourth Assessment Report, and the President of the International Association of Cryospheric Sciences.

Nathan Bindoff
Nathaniel Bindoff is Professor of Physical Oceanography at the ACE CRC and IMAS, University of Tasmania, Australia, and a coordinating lead author of the IPCC Fourth Assessment and Fifth Assessment Reports.

Robert Bindschadler
Robert Bindschadler is Chief Scientist of the Laboratory for Hydrospheric and Biospheric Processes at NASA Goddard Space Flight Center, USA, a Senior Fellow of NASA Goddard, an AGU Fellow, and past President of the International Glaciological Society.

Peter Cox
Peter Cox is Professor and Met Office Chair in Climate System Dynamics at the University of Exeter, UK, and a lead author of the IPCC Fourth Assessment Report.

Nathalie de Noblet
Nathalie de Noblet is Research Scientist at the Laboratoire des Sciences du Climat et de l'Environnement (LSCE), Gif-sur-Yvette, France.

Matthew England
Matthew England is an Australian Research Council Laureate Fellow, Professor of Physical Oceanography, and joint director of the UNSW Climate Change Research Centre (CCRC) at the University of New South Wales, Australia.

Jane Francis
Jane Francis is Professor of Palaeoclimatology at the University of Leeds, UK, and Director of the Leeds Centre for Polar Science.

Nicolas Gruber
Nicolas Gruber is Professor of Environmental Physics at ETH Zurich, Switzerland, and a contributing author of the IPCC Fourth Assessment Report.

Alan Haywood
Alan Haywood is Reader in Palaeoclimatology at the School of Earth and Environment, University of Leeds, UK, and a recent recipient of the Philip Leverhulme Prize.

David Karoly
David Karoly is Professor of Meteorology and an ARC Federation Fellow at the University of Melbourne, Australia, and a lead author of the IPCC Third and Fourth Assessment Reports.

Georg Kaser
Georg Kaser is a glaciologist at the University of Innsbruck, Austria, a lead author of the IPCC Fourth Assessment Report and the IPCC Technical Paper on Climate Change and Water, and the Immediate Past President of the International Association of Cryospheric Sciences.

Corinne Le Quéré
Corinne Le Quéré is Professor of Climate Change Science and Policy, Director of the Tyndall Centre for Climate Change Research, co-Chair of the Global Carbon Project, and a lead author of the IPCC Third, Fourth, and Fifth Assessment Reports.

Tim Lenton
Tim Lenton is Professor of Earth System Science at the University of East Anglia, UK, and the recipient of the Times Higher Education Award for Research Project of the Year 2008 for his work on climate tipping points.

Michael Mann
Michael E. Mann is a Professor in the Department of Meteorology at Penn State University, USA, Director of the Penn State Earth System Science Center, and a lead author of the IPCC Third Assessment Report.

Ben McNeil
Ben McNeil is an Australian Research Council Queen Elizabeth II Research Fellow at the Climate Change Research Centre at the University of New South Wales, Australia, and an expert reviewer of the IPCC Fourth Assessment Report.

Andy Pitman

Andy Pitman is the Director of the ARC Centre of Excellence for Climate System Science at the University of New South Wales, Australia, a lead author of the IPCC Third and Fourth Assessment Reports, and a review editor in the Fifth Assessment Report.

Stefan Rahmstorf

Stefan Rahmstorf is Professor of Physics of the Oceans and department head at the Potsdam Institute for Climate Impact Research in Germany, a lead author of the IPCC Fourth Assessment Report, and a member of the German government's Advisory Council on Global Change.

Eric Rignot

Eric Rignot is a glaciologist and Senior Research Scientist at NASA's Jet Propulsion Laboratory, USA, a Professor of Earth System Science at the University of California Irvine, and a lead author of the IPCC Fourth Assessment Report.

Hans Joachim Schellnhuber

Hans Joachim Schellnhuber is Professor for Theoretical Physics and Director of the Potsdam Institute for Climate Impact Research (PIK), Chair of the German Advisory Council on Global Change (WBGU), External Professor at the Santa Fe Institute and member of the High-Level Expert Group on Energy & Climate Change advising J.M. Barroso, the President of the European Commission.

Stephen Schneider

Stephen Schneider was the Lane Professor of Interdisciplinary Environmental Studies at Stanford University, an IPCC lead author of all four Assessment and two Synthesis Reports, and founder and editor of the journal *Climatic Change*.

Steven Sherwood

Steven Sherwood is Professor of atmospheric sciences at the Climate Change Research Centre at the University of New South Wales, Australia, and a contributing author to the IPCC Fourth Assessment Report.

Richard Somerville

Richard C. J. Somerville is Distinguished Professor Emeritus at Scripps Institution of Oceanography, University of California, San Diego, USA, and a coordinating lead author of the IPCC Fourth Assessment Report.

Konrad Steffen

Konrad Steffen is Director of the Cooperative Institute for Research in Environmental Sciences (CIRES) and Professor of Climatology at the University of Colorado in Boulder, USA, and the Chair of the World Climate Research Programme's Climate and Cryosphere (CliC) project.

Eric Steig

Eric J. Steig is Director of the Quaternary Research Center and Professor of Earth and Space Sciences at the University of Washington, USA.

Martin Visbeck

Martin Visbeck is Professor of Physical Oceanography and Deputy Director of the Leibniz Institute of Marine Sciences, IFM-GEOMAR, Germany, Chair of Kiel's multidisciplinary research cluster of excellence "The Future Ocean," and co-Chair of the World Climate Research Programme's Climate Variability and Predictability (CLIVAR) Project.

Andrew Weaver

Andrew Weaver is Professor and Canada Research Chair in Climate Modelling and Analysis at the University of Victoria, Canada, and a Lead Author of the IPCC Second, Third, Fourth and Fifth Assessment Reports. He was Chief Editor of the Journal of Climate from 2005-2009.